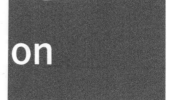

ASE on

Automotive Technician Certification Series

Engine Performance (A8)

SEP 1 7 2018

5th Edition

DISCARD

DELMAR
CENGAGE Learning

Australia • Brazil • Japan • Korea • Mexico • Singapore • Spain • United Kingdom • United States

DELMAR
CENGAGE Learning™

ASE Test Preparation: Automotive Technician Certification Series, Engine Performance (A8), 5th Edition

Vice President, Technology and Trades Professional Business Unit: Gregory L. Clayton

Director of Building Trades and transportation Training: Taryn Zlatin McKenzie

Product Manager: Katie McGuire

Associate Content Developer: Jennifer Wheaton

Director of Marketing: Beth A. Lutz

Marketing Manager: Jennifer Barbic

Senior Production Director: Wendy Troeger

Production Manager: Sherondra Thedford

Content Project Manager: PreMediaGlobal

Senior Art Director: Benj Gleeksman

Section Opener Image: Image Copyright SteveMann, 2012. Used under license from Shutterstock.com

For product information and technology assistance, contact us at
Cengage Learning Customer & Sales Support, 1-800-354-9706

For permission to use material from this text or product, submit all requests online at **www.cengage.com/permissions.**
Further permissions questions can be e-mailed to
permissionrequest@cengage.com

ISBN-13: 978-1-111-12710-7

ISBN-10: 1-111-12710-7

Delmar Cengage Learning
5 Maxwell Drive
Clifton Park, NY 12065-2919
USA

Cengage Learning is a leading provider of customized learning solutions with office locations around the globe, including Singapore, the United Kingdom, Australia, Mexico, Brazil, and Japan. Locate your local office at: **international.cengage.com/region**.

Cengage Learning products are represented in Canada by Nelson Education, Ltd.

For more information on transportation titles available from Delmar, Cengage Learning, please visit our website at **www.trainingbay.cengage.com**.

Visit our corporate website at **www.cengage.com**.

Notice to the Reader

Publisher does not warrant or guarantee any of the products described herein or perform any independent analysis in connection with any of the product information contained herein. Publisher does not assume, and expressly disclaims, any obligation to obtain and include information other than that provided to it by the manufacturer. The reader is expressly warned to consider and adopt all safety precautions that might be indicated by the activities described herein and to avoid all potential hazards. By following the instructions contained herein, the reader willingly assumes all risks in connection with such instructions. The publisher makes no representations or warranties of any kind, including but not limited to, the warranties of fitness for particular purpose or merchantability, nor are any such representations implied with respect to the material set forth herein, and the publisher takes no responsibility with respect to such material. The publisher shall not be liable for any special, consequential, or exemplary damages resulting, in whole or in part, from the readers' use of, or reliance upon, this material.

Table of Contents

Preface v

SECTION 1 **The History and Purpose of ASE** 1

SECTION 2 **Overview and Introduction** 2
Exam Administration 2
Understanding Test Question Basics 2
Test-Taking Strategies 3
Preparing for the Exam 4
What to Expect During the Exam 5
Testing Time 5
Understanding How Your Exam Is Scored 6

SECTION 3 **Types of Questions on an ASE Exam** 8
Multiple-Choice/Direct Questions 8
Completion Questions 9
Technician A, Technician B Questions 10
EXCEPT Questions 10
LEAST LIKELY Questions 11
Summary 12

SECTION 4 **Task List Overview** 13
Introduction 13
Engine Performance (Test A8) Task List 14

SECTION 5 **Sample Preparation Exams** 44
Introduction 44
Preparation Exam 1 44

Preparation Exam 2 55

Preparation Exam 3 65

Preparation Exam 4 74

Preparation Exam 5 85

Preparation Exam 6 95

SECTION 6 **Answer Keys and Explanations 104**

Introduction 104

Preparation Exam 1—Answer Key 104

Preparation Exam 1—Explanations 105

Preparation Exam 2—Answer Key 123

Preparation Exam 2—Explanations 123

Preparation Exam 3—Answer Key 142

Preparation Exam 3—Explanations 143

Preparation Exam 4—Answer Key 161

Preparation Exam 4—Explanations 161

Preparation Exam 5—Answer Key 181

Preparation Exam 5—Explanations 182

Preparation Exam 6—Answer Key 201

Preparation Exam 6—Explanations 202

SECTION 7 **Appendices 220**

Preparation Exam Answer Sheet Forms 220

Glossary 226

Delmar, a part of Cengage Learning, is very pleased that you have chosen to use our ASE Test Preparation Guide to help prepare yourself for the Engine Performance (A8) ASE certification examination. This guide is designed to help prepare you for your actual exam by providing you with an overview and introduction of the testing process, introducing you to the task list for the Engine Performance (A8) certification exam, giving you an understanding of what knowledge and skills you are expected to have in order to successfully perform the duties associated with each task area, and providing you with several preparation exams designed to emulate the live exam content in hopes of assessing your overall exam readiness.

If you have a basic working knowledge of the discipline you are testing for, you will find this book is an excellent guide, helping you understand the "must know" items needed to successfully pass the ASE certification exam. This manual is not a textbook. Its objective is to prepare the individual who has the existing requisite experience and knowledge to attempt the challenge of the ASE certification process. This guide cannot replace the hands-on experience and theoretical knowledge required by ASE to master the vehicle repair technology associated with this exam. If you are unable to understand more than a few of the preparation questions and their corresponding explanations in this book, it could be that you require either more shop-floor experience or further study.

This book begins by providing an overview of, and introduction to, the testing process. This section outlines what we recommend you do to prepare, what to expect on the actual test day, and overall methodologies for your success. This section is followed by a detailed overview of the ASE task list to include explanations of the knowledge and skills you must possess to successfully answer questions related to each particular task. After the task list, we provide six sample preparation exams for you to use as a means of evaluating areas of understanding, as well as areas requiring improvement in order to successfully pass the ASE exam. Delmar is the first and only test preparation organization to provide so many unique preparation exams. We enhanced our guides to include this support as a means of providing you with the best preparation product available. Section 6 of this guide includes the answer keys for each preparation exam, along with the answer explanations for each question. Each answer explanation also contains a reference back to the related task or tasks that it assesses. This will provide you with a quick and easy method for referring back to the task list whenever needed. The last section of this book contains blank answer sheet forms you can use as you attempt each preparation exam, along with a glossary of terms.

OUR COMMITMENT TO EXCELLENCE

Thank you for choosing Delmar, Cengage Learning for your ASE test preparation needs. All of the writers, editors, and Delmar staff have worked very hard to make this test preparation guide second to none. We feel confident that you will find this guide easy to use and extremely beneficial as you prepare for your actual ASE exam.

Delmar, Cengage Learning has sought out the best subject matter experts in the country to help with the development of *ASE Test Preparation: Automotive Technician Certification Series, Engine Performance (A8), 5th Edition*. Preparation questions are authored and then reviewed by a group of certified, subject-matter experts to ensure the highest level of quality and validity to our product.

If you have any questions concerning this guide or any guide in this series, please visit us on the web at **http://www.trainingbay.cengage.com**.

For web-based online test preparation for ASE certifications; please visit us on the web at **http://www.techniciantestprep.com/** to learn more.

ABOUT THE AUTHOR

Doug Poteet has been around all types of vehicles his entire life. He was raised on a small farm in central Kentucky, where his father, Gordon, got him interested in fixing cars, trucks, and tractors at a young age. Doug earned his diploma from Nashville Auto/Diesel College at age 17 and went on to become a heavy truck technician; after several years as a truck technician, Doug became a line technician at a new car dealership. For the next 15 years, he specialized in drivability. In 1994, he joined the faculty at Elizabethtown Community and Technical College, where he is currently serves as Associate Professor for the Automotive and Diesel programs. He earned an Associate of Science degree in Vocational/Technical Education from Western Kentucky University in Bowling Green. Doug holds ASE certifications in Master Automotive Technician, Master Medium/Heavy Truck Technician, School Bus Technician, Advanced Engine Performance, and Parts Specialist. In his free time, Doug enjoys hunting, climbing hills in his rail buggy with his wife, Sandy, and spending time with his grown sons, Kirk and Wesley.

ABOUT THE SERIES ADVISOR

Mike Swaim has been an Automotive Technology Instructor at North Idaho College, Coeur d'Alene, Idaho since 1978. He is an Automotive Service Excellence (ASE) Certified Master Technician since 1974 and holds a Lifetime Certification from Mobile Air Conditioning Society. He served as Series Advisor to all nine of the 2011 Automotive Technician/Light Truck Technician Certification Tests (A Series) of Delmar, Cengage ASE Test Preparation titles, and is the author of *ASE Test Preparation: Automotive Technician Certification Series, Undercar Specialist Designation (X1), 5th Edition.*

The History and Purpose of ASE

ASE began as the National Institute for Automotive Service Excellence (NIASE). It was founded as a non-profit, independent entity in 1972 by a group of industry leaders with the single goal of providing a means for consumers to distinguish between incompetent and competent technicians. It accomplishes this goal through the testing and certification of repair and service professionals. Though it is still known as the National Institute for Automotive Service Excellence, it is now called "ASE" for short.

Today, ASE offers more than 40 certification exams in automotive, medium/heavy duty trucks, collision repair and refinish, school bus, transit bus, parts specialist, automobile service consultant, and other industry-related areas. At this time, there are more than 385,000 professionals nationwide with current ASE certifications. These professionals are employed by new car and truck dealerships, independent repair facilities, fleets, service stations, franchised service facilities, and more.

ASE's certification exams are industry-driven and cover practically every on-highway vehicle service segment. The exams are designed to stress the knowledge of job-related skills. Certification consists of passing at least one exam and documenting two years of relevant work experience. To maintain certification, those with ASE credentials must be re-tested every five years.

While ASE certifications are a targeted means of acknowledging the skills and abilities of an individual technician, ASE also has a program designed to provide recognition for highly qualified repair, support, and parts businesses. The Blue Seal of Excellence Recognition Program, allows businesses to showcase their technicians and their commitment to excellence. One of the requirements of becoming Blue Seal recognized is that the facility must have a minimum of 75 percent of their technicians ASE certified. Additional criteria apply, and program details can be found on the ASE website.

ASE recognized that educational programs serving the service and repair industry also needed a way to be recognized as having the faculty, facilities, and equipment to provide quality education to students wanting to become service professionals. Through the combined efforts of ASE, industry, and education leaders, the non-profit organization entitled the National Automotive Technicians Education Foundation (NATEF) was created in 1983 to evaluate and recognize academic programs. Today more than 2,000 educational programs are NATEF certified.

For additional information about ASE, NATEF, or any of their programs, the following contact information can be used:

National Institute for Automotive Service Excellence (ASE)

101 Blue Seal Drive S.E.

Suite 101

Leesburg, VA 20175

Telephone: 703-669-6600

Fax: 703-669-6123

Website: **www.ase.com**

Overview and Introduction

Participating in the National Institute for Automotive Service Excellence (ASE) voluntary certification program provides you with the opportunity to demonstrate you are a qualified and skilled professional technician who has the "know-how" required to successfully work on today's modern vehicles.

EXAM ADMINISTRATION

> *Note:* After November 2011, ASE will no longer offer paper and pencil certification exams. There will be no Winter testing window in 2012, and ASE will offer and support CBT testing exclusively starting in April 2012.

ASE provides computer-based testing (CBT) exams, which are administered at test centers across the nation. It is recommended that you go to the ASE website at http://www.ase.com and review the conditions and requirements for this type of exam. There is also an exam demonstration page that allows you to personally experience how this type of exam operates before you register.

CBT exams are available four times annually, for two-month windows, with a month of no testing in between each testing window:

- January/February – Winter testing window
- April/May – Spring testing window
- July/August – Summer testing window
- October/November – Fall testing window

Please note, testing windows and timing may change. It is recommended you go to the ASE website at *http://www.ase.com* and review the latest testing schedules.

UNDERSTANDING TEST QUESTION BASICS

ASE exam questions are written by service industry experts. Each question on an exam is created during an ASE-hosted "item-writing" workshop. During these workshops, expert service representatives from manufacturers (domestic and import), aftermarket parts and equipment manufacturers, working technicians, and technical educators gather to share ideas and convert them into actual exam questions. Each exam question written by these experts must then survive review by all members of the group. The questions are designed to address the practical application of repair and diagnosis knowledge and skills practiced by technicians in their day-to-day work.

After the item-writing workshop, all questions are pre-tested and quality-checked on a national sample of technicians. Those questions that meet ASE standards of quality and accuracy are included in the scored sections of the exams; the "rejects" are sent back to the drawing board or discarded altogether.

Depending on the topic of the certification exam, you will be asked between 40 and 80 multiple-choice questions. You can determine the approximated number of questions you can expect to be asked during the Engine Performance (A8) certification exam by reviewing the task list in Section 4 of this book. The five-year recertification exam will cover this same content; however, the number of questions for each content area of the recertification exam will be reduced by approximately one-half.

> *Note:* Exams may contain questions that are included for statistical research purposes only. Your answers to these questions will not affect your score, but since you do not know which ones they are, you should answer all questions in the exam.

Using multiple criteria, including cross-sections by age, race, and other background information, ASE is able to guarantee that exam questions do not include bias for or against any particular group. A question that shows bias toward any particular group is discarded.

TEST-TAKING STRATEGIES

Before beginning your exam, quickly look over the exam to determine the total number of questions that you will need to answer. Having this knowledge will help you manage your time throughout the exam to ensure you have enough available to answer all of the questions presented. Read through each question completely before marking your answer. Answer the questions in the order they appear on the exam. Leave the questions blank that you are not sure of and move on to the next question. You can return to those unanswered questions after you have finished the others. These questions may actually be easier to answer at a later time, once your mind has had additional time to consider them on a subconscious level. In addition, you might find information in other questions that will help you recall the answers to some of them.

Multiple-choice exams are sometimes challenging because there are often several choices that may seem possible, or partially correct, and therefore it may be difficult to decide on the most appropriate answer choice. The best strategy, in this case, is to first determine the correct answer before looking at the answer options. If you see the answer you decided on, you should still be careful to examine the other answer options to make sure that none seems more correct than yours. If you do not know or are not sure of the answer, read each option very carefully and try to eliminate those options that you know are incorrect. That way, you can often arrive at the correct choice through a process of elimination.

If you have gone through the entire exam, and you still do not know the answer to some of the questions, *then guess*. Yes, guess. You then have at least a 25 percent chance of being correct. While your score is based on the number of questions answered correctly, any question left blank, or unanswered, is automatically scored as incorrect.

There is a lot of "folk" wisdom on the subject of test taking that you may hear about as you prepare for your ASE exam. For example, there are those who would advise you to avoid response options that use certain words such as *all, none, always, never, must,* and *only,* to name a few. This, they claim, is because nothing in life is exclusive. They would advise you to choose response options that use words that allow for some exception, such as *sometimes, frequently, rarely, often, usually, seldom,* and *normally.* They would also advise you to avoid the first and last option (A or D) because exam writers, they feel, are more comfortable if they put the correct answer in the middle (B or C) of the choices. Another recommendation often offered is to select the option that is either shorter or longer than the other three choices because it is more likely to be correct. Some would advise you to never change an answer since your first intuition is usually correct. Another area of "folk" wisdom focuses specifically on any repetitive patterns created by your question responses (e.g., A, B, C, A, B, C, A, B, C).

Many individuals may say that there are actual grains of truth in this "folk" wisdom, and whereas with some exams, this may prove true, it is not relevant in regard to the ASE certification exams. ASE validates all exam questions and test forms through a national sample of technicians, and only those questions and test forms that meet ASE standards of quality and accuracy are included in the scored sections of the exams. Any biased questions or patterns are discarded altogether, and therefore, it is highly unlikely you will experience any of this "folk" wisdom on an actual ASE exam.

PREPARING FOR THE EXAM

Delmar, Cengage Learning wants to make sure we are providing you with the most thorough preparation guide possible. To demonstrate this, we have included hundreds of preparation questions in this guide. These questions are designed to provide as many opportunities as possible to prepare you to successfully pass your ASE exam. The preparation approach we recommend and outline in this book is designed to help you build confidence in demonstrating what task area content you already know well while also outlining what areas you should review in more detail prior to the actual exam.

We recommend that your first step in the preparation process should be to thoroughly review Section 3 of this book. This section contains a description and explanation of the type of questions you'll find on an ASE exam.

Once you understand how the questions will be presented, we then recommend that you thoroughly review Section 4 of this book. This section contains information that will help you establish an understanding of what the exam will be evaluating, and specifically, how many questions to expect in each specific task area.

As your third preparatory step, we recommend you complete your first preparation exam, located in Section 5 of this book. Answer one question at a time. After you answer each question, review the answer and question explanation information located in Section 6. This section will provide you with instant response feedback, allowing you to gauge your progress, one question at a time, throughout this first preparation exam. If after reading the question explanation you do not feel you understand the reasoning for the correct answer, go back and review the task list overview (Section 4) for the task that is related to that question. Included with each question explanation is a clear identifier of the task area that is being assessed (e.g., Task A.1). If at that point you still do not feel you have a solid understanding of the material, identify a good source of information on the topic, such as an educational course, textbook, or other related source of topical learning, and do some additional studying.

After you have completed your first preparation exam and have reviewed your answers, you are ready to complete your next preparation exam. A total of six practice exams are available in Section 5 of this book. For your second preparation exam, we recommend that you answer the questions as if you were taking the actual exam. Do not use any reference material or allow any interruptions in order to get a feel for how you will do on the actual exam. Once you have answered all of the questions, grade your results using the Answer Key in Section 6. For every question that you gave an incorrect answer to, study the explanations to the answers and/or the overview of the related task areas. Try to determine the root cause for missing the question. The easiest thing to correct is learning the correct technical content. The hardest things to correct are behaviors that lead you to an incorrect conclusion. If you knew the information but still got the question incorrect, there is likely a test-taking behavior that will need to be corrected. An example of this would be reading too quickly and skipping over words that affect your reasoning. If you can identify what you did that caused you to answer the question incorrectly, you can eliminate that cause and improve your score.

Here are some basic guidelines to follow while preparing for the exam:

- Focus your studies on those areas in which you are weak.
- Be honest with yourself when determining if you understand something.
- Study often but for short periods of time.
- Remove yourself from all distractions when studying.
- Keep in mind that the goal of studying is not just to pass the exam; the real goal is to learn.
- Prepare physically by getting a good night's rest before the exam, and eat meals that provide energy but do not cause discomfort.
- Arrive early to the exam site to avoid long waits as test candidates check in.
- Use all of the time available for your exams. If you finish early, spend the remaining time reviewing your answers.
- Do not leave any questions unanswered. If absolutely necessary, guess. All unanswered questions are automatically scored as incorrect.

Here are some items you will need to bring with you to the exam site:

- A valid government or school-issued photo ID
- Your test center admissions ticket
- A watch (not all test sites have clocks)

> *Note:* Books, calculators, and other reference materials are not allowed in the exam room. The exceptions to this list are English-Foreign dictionaries, or glossaries. All items will be inspected before and after testing.

WHAT TO EXPECT DURING THE EXAM

When taking a CBT exam, as soon as you are seated in the testing center, you will be given a brief tutorial to acquaint you with the computer-delivered test, prior to taking your certification exam(s). The CBT exams allow you to select only one answer per question. You can also change your answers as many times as you like. When you select a second answer choice, the CBT will automatically unselect your first answer choice. If you want to skip a question to return to later, you can utilize the "flag" feature, which will allow you to quickly identify and review questions whenever you are ready. Prior to completing your exam, you will also be provided with an opportunity to review your answers and address any unanswered questions.

TESTING TIME

Individual ASE CBT exam has a fixed time limit. Individual exam times will vary based upon exam area and will range anywhere from a half hour to two hours. You will also be given an additional 30 minutes beyond what is allotted to complete your exams to ensure you have adequate time to perform all necessary check-in procedures, complete a brief CBT tutorial, and potentially complete a post-test survey.

You can register for and take multiple CBT exams during one testing appointment. The maximum time allotment for a CBT appointment is four and a half hours. If you happen to register for so many exams that you will require more time than this, your exams will be scheduled into multiple appointments. This could mean that you have testing on both the morning and afternoon of the same day, or they could be scheduled on different days, depending on your personal preference and the test center's schedule.

It is important to understand that if you arrive late for your CBT test appointment, you will not be able to make up any missed time. You will only have the scheduled amount of time remaining in your appointment to complete your exam(s).

Also, while most people finish their CBT exams within the time allowed, others might feel rushed or not be able to finish the test, due to the implied stress of a specific, individual time limit allotment. Before you register for the CBT exams, you should review the number of exam questions that will be asked along with the amount of time allotted for that exam to determine whether you feel comfortable with the designated time limitation or not.

As an overall time management recommendation, you should monitor your progress and set a time limit you will follow with regard to how much time you will spend on each individual exam question. This should be based on the total number of questions you will be answering.

Also, it is very important to note that if for any reason you wish to leave the testing room during an exam, you must first ask permission. If you happen to finish your exam(s) early and wish to leave the testing site before your designated session appointment is completed, you are permitted to do so only during specified dismissal periods.

UNDERSTANDING HOW YOUR EXAM IS SCORED

You can gain a better perspective about the ASE certification exams if you understand how they are scored. ASE exams are scored by an independent organization having no vested interest in ASE or in the automotive industry.

Each question carries the same weight as any other question. For example, if there are 50 questions, each is worth 2 percent of the total score.

Your exam results can tell you:

- Where your knowledge equals or exceeds that needed for competent performance, or
- Where you might need more preparation.

Your ASE exam score report is divided into content "task" areas; it will show the number of questions in each content area and how many of your answers were correct. These numbers provide information about your performance in each area of the exam. However, because there may be a different number of questions in each content area of the exam, a high percentage of correct answers in an area with few questions may not offset a low percentage in an area with many questions.

It should be noted that one does not "fail" an ASE exam. The technician who does not pass is simply told "More Preparation Needed." Though large differences in percentages may indicate problem areas, it is important to consider how many questions were asked in each area. Since each exam evaluates all phases of the work involved in a service specialty, you should be prepared in each area. A low score in one area could keep you from passing an entire exam. If you do not pass the exam, you may take it again at any time it is scheduled to be administered.

There is no such thing as average. You cannot determine your overall exam score by adding the percentages given for each task area and dividing by the number of areas. It does not work that way because there generally is not the same number of questions in each task area. A task area with 20 questions, for example, counts more toward your total score than a task area with 10 questions.

Your exam report should give you a good picture of your results and a better understanding of your strengths and areas needing improvement for each task area.

Types of Questions on an ASE Exam

Understanding not only what content areas will be assessed during your exam, but how you can expect exam questions to be presented will enable you to gain the confidence you need to successfully pass an ASE certification exam. The following examples will help you recognize the types of question styles used in ASE exams and assist you in avoiding common errors when answering them.

Most initial certification tests are made up of between 40 and 80 multiple-choice questions. The five-year recertification exams will cover the same content as the initial exam; however, the actual number of questions for each content area will be reduced by approximately one-half. Refer to Section 4 of this book for specific details regarding the number of questions to expect during the initial Engine Performance (A8) certification exam.

Multiple-choice questions are an efficient way to test knowledge. To correctly answer them, you must consider each answer choice as a possibility, and then choose the answer choice that *best* addresses the question. To do this, read each word of the question carefully. Do not assume you know what the question is about until you have finished reading the entire question.

About 10 percent of the questions on an actual ASE exam will reference an illustration. These drawings contain the information needed to correctly answer the question. The illustration should be studied carefully before attempting to answer the question. When the illustration is showing a system in detail, look over the system and try to figure out how the system works before you look at the question and the possible answers. This approach will ensure that you do not answer the question based upon false assumptions or partial data, but instead have reviewed the entire scenario being presented.

MULTIPLE-CHOICE/DIRECT QUESTIONS

The most common type of question used on an ASE exam is the direct multiple-choice style question. This type of question contains an introductory statement, called a stem, followed by four options: three incorrect answers, called distracters, and one correct answer, the key. When the questions are written, the point is to make the distracters plausible to draw an inexperienced technician to inadvertently select one of them. This type of question gives a clear indication of the technician's knowledge.

Here is an example of a direct style question:

TASK A.11

1. Which of the following would be used to measure valve lash clearance on a gasoline engine?

   A. Dial indicator

   B. Feeler gauge

   C. Plastigauge®

   D. Dial caliper

Answer A is incorrect. A dial indicator is used to measure end-play or runout. It would not be useful to measure valve lash clearance.

Answer B is correct. A feeler gauge is used for measuring valve lash clearance.

Answer C is incorrect. Plastigauge® is the most common method of measuring crankshaft main bearing clearance.

Answer D is incorrect. A dial caliper is used to measure outside depth or inside dimensions, but would not be an effective tool to measure valve lash clearance.

2. A vehicle with electronic throttle control has an accelerator pedal position diagnostic trouble code. The technician wiggles the wiring harness while observing the accelerator pedal position sensor voltage with the scan tool. The sensor voltage changes. Which of the following is the most likely cause of the diagnostic trouble code?

TASK A.1

   A. Faulty scan tool
   B. Faulty electronic control module (ECM)
   C. Faulty accelerator pedal position sensor wiring
   D. Faulty ECM power supply

Answer A is incorrect. If the voltage value changed while moving the wiring harness, there is no reason to believe the scan tool is faulty.

Answer B is incorrect. If the voltage value changed while moving the wiring harness, there is no reason to believe the ECM is faulty.

Answer C is correct. If the voltage value changed while moving the wiring harness, the most likely cause is the wiring harness.

Answer D is incorrect. If all other items associated with the ECM are normal and the voltage value changed while moving the wiring harness, the most likely cause is the wiring harness.

COMPLETION QUESTIONS

A completion question is similar to the direct question except the statement may be completed by any one of the four options to form a complete sentence.

Here is an example of a completion question:

1. A technician will most likely refer to a wiring diagram to find:

   A. The power and ground distribution for the circuit
   B. The location of the ground connection
   C. Updated factory information about pattern failures
   D. A flowchart for troubleshooting an electrical problem

Answer A is correct. A wiring diagram will provide details about how the power and ground are connected to the circuit.

Answer B is incorrect. A wiring diagram does not typically provide the location of electrical components such as a ground connection.

Answer C is incorrect. A wiring diagram does not typically provide updated factory information. Technical service bulletins provide updated factory information about pattern failures.

Answer D is incorrect. A wiring diagram does not provide any flowcharts for troubleshooting.

TECHNICIAN A, TECHNICIAN B QUESTIONS

This type of question is usually associated with an ASE exam. It is, in fact, two true-false statements grouped together, such as: "Technician A says ..." and "Technician B says ...", followed by "Who is correct?

In this type of question, you must determine whether either, both, or neither of the statements are correct. To answer this type of question correctly, you must carefully read each technician's statement and judge it on its own merit.

Sometimes this type of question begins with a statement about some analysis or repair procedure. This statement provides the setup or background information required to understand the conditions about which Technician A and Technician B are talking, followed by two statements about the cause of the concern, proper inspection, identification, or repair choices. Analyzing this type of question is a little easier than the other types because there are only two ideas to consider, although there are still four choices for an answer.

Again, Technician A, Technician B questions are really double true-or-false questions. The best way to analyze this type of question is to consider each technician's statement separately. Ask yourself, "Is A true or false? Is B true or false?" Once you have completed an individual evaluation of each statement, you will have successfully determined the correct answer choice for the question, "Who is correct?" An important point to remember is that an ASE Technician A, Technician B question will never have Technician A and B directly disagreeing with each other. That is why you must evaluate each statement independently.

An example of a Technician A/Technician B style question looks like this:

TASK A.21

1. A noise is heard from the accessory drive on the front of a gasoline engine. Technician A says that the serpentine belt can be removed to help determine if it is the source of the noise. Technician B says water dripped on the belt can help determine if the belt is the source of the noise. Who is correct?

    A. A only
    B. B only
    C. Both A and B
    D. Neither A nor B

Answer A is incorrect. Technician B is also correct.

Answer B is incorrect. Technician A is also correct.

Answer C is correct. Both technicians are correct. The belt can be removed and the engine run for a short time to determine if the belt is the source of the noise. Also, if the noise disappears when a few drops of water are put on the running belt, the technician knows the belt is the source of the noise.

Answer D is incorrect. Both technicians are correct.

EXCEPT QUESTIONS

Another type of question type used on ASE exams contains answer choices that are all correct except for one. To help easily identify this type of question, whenever it is presented in an exam, the word "EXCEPT" will always be displayed in capital letters.

Be careful to read these question types slowly and thoroughly; otherwise, you may overlook what the question is actually asking and answer the question by selecting the first correct statement. Furthermore, a cautionary statement will alert you to the fact that the next question is different from the ones otherwise found in the exam. With the EXCEPT type of question, only one incorrect choice will actually be listed among the options, and that incorrect choice will be the key to the question. That is, the incorrect statement is counted as the correct answer for that question.

An example of this type of question would appear as follows:

1. A gasoline engine is being checked for an engine oil leak. All of the following could be used to help locate the source of the leak EXCEPT:

    A.   Blacklight

    B.   White powder

    C.   Vacuum gauge

    D.   Oil dye

TASK A.6

Answer A is incorrect. A blacklight can be used to help locate the source of a leak.

Answer B is incorrect. White powder can be used to help locate the source of a leak.

Answer C is correct. A vacuum gauge is not used to locate engine oil leaks.

Answer D is incorrect. Oil dye can be used to help locate an engine oil leak.

LEAST LIKELY QUESTIONS

LEAST LIKELY questions are similar to EXCEPT questions. Look for the answer choice that would be the LEAST LIKELY cause (most incorrect) of the described situation. To help easily identify these type of questions, whenever they are presented in an exam, the words "LEAST LIKELY" will always be displayed in capital letters. In addition, you will be alerted before a LEAST LIKELY question is posed. Read the entire question carefully before choosing your answer.

An example of this type of question is shown below:

1. Which of the following is the LEAST LIKELY cause of ignition module failure?

    A.   An open spark plug wire

    B.   No dielectric grease under the module

    C.   A fouled spark plug

    D.   Loose module mounting screws

TASK B.7

Answer A is incorrect. Faulty secondary ignition components are common causes of ignition module failure.

Answer B is incorrect. Some modules require the use of dielectric grease to aid in dissipating heat.

Answer C is correct. A fouled spark plug would cause an engine miss, but would not cause the ignition module to fail.

Answer D is incorrect. Loose module mounting screws will cause the module to fail prematurely.

SUMMARY

The question styles outlined in this section are the only ones you will encounter on any ASE certification exam. ASE does not use any other types of question styles, such as fill-in-the-blank, true/false, word-matching, or essay. ASE also will not require you to draw diagrams or sketches to support any of your answer selections, although any of the described question styles may include illustrations, charts, or schematics to clarify a question. If a formula or chart is required to answer a question, it will be provided for you.

Task List Overview

INTRODUCTION

This section of the book outlines the content areas, or *task list*, for this specific certification exam, along with a written overview of the content covered in the exam.

The task list describes the actual knowledge and skills necessary for a technician to successfully perform the work associated with each skill area. This task list is the fundamental guideline you should use to understand what areas you can expect to be tested on, as well as how each individual area is weighted to include the approximate number of questions you can expect to be given for that area during the ASE certification exam. It is important to note that the number of exam questions for a particular area is to be used only as a guideline. ASE advises that the questions on the exam may not equal the number specifically listed on the task list. The task lists are specifically designed to tell you what ASE expects you to know how to do and to help you get ready to be tested.

Similar to the role this task list will play with regard to the actual ASE exam, Delmar, Cengage Learning has developed the six preparation exams, located in Section 5 of this book, using this task list as a guide. It is important to note that both ASE and Delmar, Cengage Learning uses the same task list as a guideline for creating these test questions. None of the test questions you will see in this book will be found in the actual, live ASE exams. This is true for any test preparatory material you use. Real exam questions are *only* visible during the actual ASE exams.

Task List at a Glance

The Engine Performance (A8) task list focuses on five core areas, and you can expect to be asked approximately 50 questions on your certification exam, broken out as outlined:

- A. General Diagnosis (12 questions)
- B. Ignition System Diagnosis and Repair (8 questions)
- C. Fuel, Air Induction, and Exhaust Systems Diagnosis and Repair (9 questions)
- D. Emissions Control Systems Diagnosis and Repair (Including OBD II) (8 questions)
    1. Positive Crankcase Ventilation (1)
    2. Exhaust Gas Recirculation (2)
    3. Secondary Air Injection (AIR) and Catalytic Converter (2)
    4. Evaporative Emissions Controls (3)
- E. Computerized Engine Controls Diagnosis and Repair (Including OBD II) (13 questions)

Based upon this information, the graph shown here is a general guideline demonstrating which areas will have the most focus on the actual certification exam. This data may help you prioritize your time when preparing for the exam.

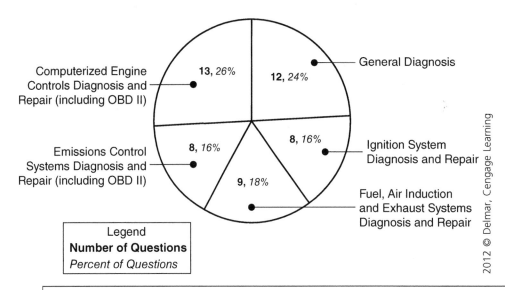

Note: The actual number of questions you will be given on the ASE certification exam may vary slightly from the information provided in the task list, as exams may contain questions that are included for statistical research purposes only. Don't forget that your answers to these research questions will not affect your score.

ENGINE PERFORMANCE (TEST A8) TASK LIST

A. General Engine Diagnosis (12 Questions)

1. Verify driver's complaint, perform visual inspection, and/or road test vehicle; determine needed action.

When customers bring their vehicle to the shop for repair, they have a concern or problem in their mind but may have difficulty expressing the problem in words. In most repair facilities, it is the service writer's responsibility to help gather information and define the problem so the technician can diagnose it.

Many vehicle concerns can be located with a thorough visual inspection. A leaking fuel pressure regulator diaphragm may cause hard starting and poor mileage complaints. Simply removing the fuel pressure regulator vacuum hose and inspecting for the presence of fuel can isolate the problem without performing detailed fuel system tests. A visual inspection should include the following points:

- Checking all air intake plumbing for cracks or loose clamps
- Checking vacuum hoses conditions
- Routing and checking spark plug wires for misrouted or damaged cables
- Checking wiring harnesses for proper routing and for damage from chafing brackets or components on the engine
- Checking all fluids for proper level and condition

A clear and detailed description of the customer's concerns must be obtained either from the customer or the service writer before diagnostics begin. A service technician should review previous service history of the vehicle supplied by the customer or located in the shop's service records to narrow the scope of his testing.

Performing a road test is part of almost all service routines. A road test allows the technician to verify a customer concern as well as to verify a successful repair. When test driving a vehicle, the technician should make note of any unusual noises, look for smoke from the exhaust, operate the vehicle through different speed and load ranges to determine engine and transmission performance, and pay attention to any steering, suspension, or braking problems that may be evident.

All road test finding should be well documented on the vehicle repair order. A thorough test drive will not only verify a concern or repair, but may also uncover other problems about which the customer was not aware. For instance, a broken motor mount on a rear-wheel drive (RWD) vehicle may allow the engine to lift on acceleration, which a skilled service technician may notice, but of which the customer could be unaware. Situations such as these occur frequently and illustrate the need to follow a careful routine when road testing a vehicle.

2. Research applicable vehicle and service information, such as engine-management system operation, vehicle service history, service precautions, technical service bulletins, and service campaigns/recalls.

Understanding the system you're working on is an important part of diagnosis. This applies to engine- and power train-management systems as well as other parts of the vehicle. This understanding includes theory of operation information that is usually found in the manufacturer's vehicle service manuals or aftermarket vehicle service manuals from publishers, such as Mitchell, Motors, or Chilton. These manuals also provide detailed repair information necessary to service vehicles today, such as wiring diagrams, component replacement procedures, adjustment procedures, and specifications and test procedures.

Previous vehicle service information should be gathered from the customer if repairs or service were performed elsewhere, as well as any repairs already performed by the shop that should be kept in a customer file or on a shop management computer. Knowing what repairs have already been performed will help eliminate duplicate services.

Further information is available in the form of manufacturer's *technical service bulletins* (TSBs) and recall notices. These may be in printed or electronic format and are available from the same sources as mentioned above. In electronic format, the bulletins can be searched with a personal computer by vehicle as well as system or symptom. The bulletin can then be printed or viewed on screen.

The Internet also provides many sources for information gathering. Vehicle and parts manufacturer websites are available and provide access to service information or the ability to purchase manuals and training material online.

Some handheld equipment available today has diagnostic information imbedded in the tools' software for easy technician access. This form of information access is very useful and well suited to scan tools and lab scopes or graphing multimeters.

Lastly, vehicle repair diagnostic hotline services are available and can be very helpful resources when working on unfamiliar vehicles or complex problems. Various sources such as parts suppliers, tool manufacturers, and independent companies provide these services.

3. Diagnose noises and/or vibration problems related to engine performance; determine needed action.

Engine defects such as worn pistons and cylinder walls, worn rings, loose piston pins, worn crankshaft bearings, a cracked flywheel or flex plate, worn camshaft lobes, and loose or worn valve train components usually produce their own identifiable noises or vibrations. Identifying when the noises and vibrations occur can be helpful in determining the faulty component. A stethoscope would be useful in determining the location of a noise.

Attention must be paid to the relation of the noise to engine speed as well as engine temperature to help pinpoint the source of the noise. For example, a problem such as piston slap may be very noticeable with a cold engine but almost disappear when the engine reaches operating temperature, while a worn connecting rod bearing gets louder as the engine warms up and the oil thins out.

4. Diagnose the cause of unusual exhaust color, odor, and sound; determine needed action.

If the engine is operating normally, then the exhaust should be colorless. A small swirl of white vapor from the tailpipe is normal in cold weather. This is vapor moisture in the exhaust and is a byproduct of combustion. If the exhaust is bluish gray, then small amounts of oil are entering the combustion chamber. If the exhaust is black, then the air/fuel mixture is too rich. If the exhaust is grayish white, then coolant may be leaking into the combustion chambers.

5. Perform engine manifold vacuum or pressure tests; determine needed action.

When a vacuum gauge is connected to the intake manifold, the gauge should show a steady reading between 17 and 22 in. Hg (44.8 and 27.6 kPa absolute) with the engine idling at sea level. Manifold vacuum readings will decrease approximately 1 in. Hg for every 1,000 feet increase in elevation above sea level. A low steady reading indicates late ignition timing. Burned or leaking valves cause a vacuum gauge fluctuation between 12 and 18 in. Hg. When the engine is accelerated and held at a steady 2,500 rpm and the vacuum reading slowly drops to a very low reading, an exhaust system restriction is indicated.

When performing vacuum tests, the technician should keep in mind the effect of the valve train on the production of vacuum. If the valve timing is not correct, then the engine will not perform as designed, and lower vacuum readings will result. Additionally, incorrect valve adjustment can reduce engine efficiency. Valve adjustment that is too tight (insufficient clearance) can cause a valve to hold open, reducing compression, and cause the valve to burn. Incorrect valve timing will change when the valves are open and closed, which can greatly affect the manifold vacuum. Late valve timing from a jumped timing chain or belt will cause very low vacuum readings. If the valve timing is very late, then the engine likely will not start.

A cranking vacuum test will show either no vacuum or possibly a gauge needle that bounces back and forth across zero, indicating positive pressure in the intake manifold. This occurs because, as the piston begins its upward travel on the compression stroke, the intake valve is still open. Cranking vacuum tests should be performed during any no-start diagnosis to ensure proper engine sealing. A normal cranking vacuum reading should be between 1 and 5 in. Hg at sea level.

6. Perform cylinder power balance test; determine needed action.

A cylinder power balance test is performed to ensure that all cylinders are contributing equally. When individual cylinders are disabled, a noticeable drop in *revolutions per minute* (RPM) is measured. By comparing the amount of RPM drop between cylinders, it can be determined which cylinder has a problem. Performing cylinder power balance testing on late-model computer-controlled engines often requires disabling the idle control system and forcing the engine control system into open loop by disconnecting the oxygen sensor to prevent the computer from compensating for the disabled cylinder.

Many newer engine control computers allow the technician to perform cylinder balance testing using a scan tool. The computer will disable the fuel injectors one at a time and command a fixed engine speed and open-loop operation during the test. This test is much easier and faster to perform. A cylinder balance test will only identify which cylinder is producing low power; it cannot pinpoint the exact cause. Further pinpoint testing will be necessary to determine if the problem is a compression, ignition, or fuel delivery issue.

7. Perform cylinder cranking compression test; interpret running compression test results; determine needed action.

The compression test checks the sealing quality of the combustion chamber. If the compression is lower than specified in one or more cylinders, then the valves and rings are suspect. A wet compression test is useful in deciding whether rings or valves are the problem. If compression comes up during the wet compression test, then rings are the most likely problem. Conversely, if compression does not come up, then leaking valves are the likely problem. When performing a compression test, the engine should be at operating temperature and the throttle should be held open in order to get a more accurate reading. A running compression test, sometimes called a dynamic compression test is performed with the engine running. The compression pressure of a running engine is lower than cranking compression pressure. For most engines the running compression pressure should be about half of the cranking compression pressure.

If it is determined that the valves are the cause of compression loss, then the valve clearance should be checked and adjusted before continuing with repair. In some cases, valve timing could cause compression loss. If this is the case, then the sound of the engine while cranking is distinctive and will lead the technician to investigate valve timing.

8. Perform cylinder leakage/leak down test; determine needed action.

The *cylinder leakage test* (leak down test) can be used to further pinpoint the problem. A regulated amount of air is introduced into the cylinder; the gauge on the tester will indicate what percentage of that pressure is leaking. A gauge reading of 0 percent indicates no leakage, while a reading of 100 percent indicates that cylinder does not hold compression. If two adjacent cylinders have excess leakage, then a head gasket problem is likely. While testing a cylinder with high leakage, you should try to find where the leakage is directed? For example, if air is heard escaping from the exhaust, then a leaking exhaust valve is indicated. Air coming from the intake would indicate a leaking intake valve. Leakage could show up in the radiator, indicating a problem with a head gasket or a cracked cylinder head. Air coming from the PCV system crankcase oil filter would indicate leakage past the rings.

9. Diagnose engine mechanical, electrical, electronic, fuel, and ignition problems with an oscilloscope, engine analyzer, and/or scan tool; determine needed action.

Today's computerized engine analyzers perform detailed system tests and can produce very accurate diagnostic printouts for the technician or customer. Automated test sequences for the cranking, charging, ignition, fuel, compression, and emissions systems are done when a complete test is performed. Most analyzers also have manual pinpoint test capabilities and can be used as standalone ignition and lab-grade oscilloscopes. By utilizing all of the test leads of a computerized engine analyzer and running a complete test, the technician can identify most problems that may be present in any of the systems listed above.

For instance, during a cranking test, the analyzer will measure cranking current per cylinder to determine relative compression, check the starting system for cranking speed and normal current draw, monitor the battery condition, check the ignition kV output, and test the fuel delivery by exhaust analysis of hydrocarbons during cranking. All of these measurements are done in 15 seconds of cranking time without the engine even starting. The remaining systems are checked via automated tests that are performed after the engine is started. Determining what is right with the engine prior to performing extensive pinpoint tests to locate a problem is the greatest strength of a computerized engine analyzer.

The multi-trace lab scope function of the analyzer can be used for detailed pinpoint testing of fuel or ignition system components such as oxygen sensors, throttle position sensors, or ignition primary triggering devices.

The handheld digital storage oscilloscope (DSO) has become a very useful tool for pinpoint testing of components on today's computer-controlled engines. The DSO can be used for signal analysis while test driving a vehicle if the fault is intermittent or if road load is required to create a problem. The DSO, coupled with a current probe, is a very powerful tool for testing motors and actuators that have winding or coil-like fuel injectors, ignition coils, and solenoids. Fuel pumps or small electric motors can also be accurately tested with a DSO and current probe. Most DSOs have waveform memory and recall capabilities. They can print waveforms to a printer or save the files to special computer programs that compile databases of good and bad waveforms. These features can also be used to verify repairs or to document tests or repairs for customers.

10. Prepare and inspect vehicle and analyzer for exhaust gas (HC, CO, CO2, and O2) exhaust gas analysis; perform tests and interpret exhaust gas readings.

Many states have emissions inspection programs that require vehicle owners to maintain their vehicles to certain standards. An *emissions analyzer* measures tailpipe emissions. Emissions analyzers require a warm-up period and certain calibration intervals. Some items that may be checked with an emissions analyzer include air/fuel mixture, cylinder misfire, catalytic converter defects, and leaking head gaskets.

A 4-gas emissions analyzer is capable of measuring hydrocarbons, carbon monoxide, oxygen, and carbon dioxide.

A 5-gas analyzer adds the ability to measure oxides of nitrogen (NOx), and these units are usually portable. Oxides of nitrogen are mostly produced under load, so a means of measuring this pollutant during driving conditions makes a portable gas analyzer necessary. Many states that have enhanced emissions test programs check for NOx levels, so portable analyzers have become almost necessary in these areas. Sample dilution is a major problem with exhaust gas analysis, and it is important that the technician checks both the vehicle's exhaust system and the analyzer sample probe and hose to verify they are free from leaks. Prior to performing gas analysis, if the vehicle has an air-injection system, then this must be disabled to allow for accurate measurements. Any outside air entering the gas analyzer will skew the readings and may cause incorrect diagnosis.

Diagnosing engine or emissions system problems with a gas analyzer requires understanding what each of the exhaust gases are and how they are produced so that when the levels are incorrect, the technician can determine the most likely cause.

Hydrocarbons (HC) are basically unburned fuel molecules. Anything that inhibits proper combustion in the engine can raise hydrocarbons. Normal engine out or pre-converter hydrocarbon levels for late-model closed-loop fuel control vehicles should be less than 300 parts per million. The catalytic converter can lower this level to near zero. Overadvanced ignition timing, an overly rich or lean air/fuel mixture, or any problem that causes a cylinder to misfire will raise HC levels. Fuel must be drawn into the cylinder to raise HC levels so that a cylinder with a malfunctioning fuel injector can misfire. Because no fuel is injected, however, the HC level will not be high.

Carbon monoxide (CO) is formed in the cylinder when there is an insufficient amount of air in the mixture for the amount of fuel present. Carbon monoxide is a rich indicator. When the air/fuel mixture is leaner than 14.7:1, the amount of CO produced will be very low. Normal engine out CO levels for late-model, closed-loop fuel control vehicles is less than .75 percent. The catalytic converter can lower this amount to almost zero.

Items that may cause high CO levels on a carbureted engine include incorrect carburetor adjustment, high float level, or dirty or plugged air bleeds in the carburetor. High CO levels on fuel-injected engines can be caused by high fuel pressure, leaking injectors, or a leaking fuel pressure regulator. Any sensor inputs that have an effect on fuel delivery on closed-loop fuel control vehicles can cause increased CO emissions if they are incorrect or out of calibration. The most common sensor problems include the oxygen sensor, coolant or air intake temperature sensors, manifold absolute pressure sensor, and mass airflow sensor, and throttle position sensor.

Oxygen (O_2) is not a harmful exhaust by-product, but is measured by gas analyzers to help determine if the mixture is too lean. Oxygen and carbon monoxide levels are equal at the stoichiometric or ideal air/fuel ratio of 14.7:1. When the mixture goes richer, CO will increase, and when the mixture goes lean, O_2 will increase. Oxygen is the lean indicator in the exhaust stream. Normal engine out oxygen levels for computer-controlled vehicles is less than 1.5 percent.

High O_2 levels can be caused by sample dilution, so the exhaust system and sample hose must be checked for leaks before condemning the engine. Any misfire can increase O_2 levels because the air drawn into the engine is not burned in the misfiring cylinder. If the O_2 level is high and the engine runs good, then look for sample dilution. If the O_2 level is high and the engine runs rough, then check for vacuum leaks or a cylinder misfire.

Carbon dioxide (CO_2) is a desirable exhaust by-product that is a combustion efficiency indicator. A mechanically sound engine should produce a minimum of 12 percent CO_2. Properly operating late-model closed-loop fuel control vehicles typically show 13.5 to 15.5 percent CO_2 readings. If the other gases are normal and CO_2 does not go above 12 percent, then look for a mechanical problem or exhaust sample dilution.

Oxides of nitrogen are formed when nitrogen and oxygen combine in the combustion chamber when temperatures increase above 2,500°F.

When oxides of nitrogen and hydrocarbons mix in the atmosphere and are exposed to sunlight, photochemical smog is created. NOx is controlled in the engine by adding exhaust gas to the incoming air charge to reduce combustion chamber temperatures below 2,500°F. This is the job of the *exhaust gas recirculation* (EGR) system. The three-way catalytic converter reduces the NOx still remaining in the exhaust steam even further. NOx levels should be below 1,000 ppm when measured with a 5-gas analyzer under road load.

When diagnosing engine problems with a 4- or 5-gas analyzer, the technician must evaluate all of the gases measured to make an accurate decision concerning what may be causing out-of-specification readings. For instance, a high HC reading does not indicate a rich air/fuel mixture, because hydrocarbons can increase from a lean misfire as well a rich running condition. By observing the CO and O_2 readings, a technician can determine if the high HC reading is a result of a rich mixture, because CO will also be high while O_2 will be very low. If the CO is low and O_2 is high, then a lean misfire or cylinder misfire is indicated.

11. Verify valve adjustment on engines with mechanical or hydraulic lifters.

Valve lifters are either mechanical (solid) or hydraulic. Solid lifters provide for a rigid connection between the camshaft and the valves. Hydraulic valve lifters provide for the same connection, but use oil to absorb the shock that results from the movement of the valve train.

Hydraulic lifters are designed to automatically compensate for the effects of engine temperature. Changes in temperature cause the valve train components to expand and contract. Solid lifters require a clearance between the parts of the valve train. This clearance allows for the expansion of the components as the engine gets hot. Periodic adjustment of this clearance must be made. Excessive clearance might cause a clicking noise. This clicking noise is also an indication of the hammering of the valve train parts against one another, which will result in reduced camshaft and lifter life.

Valve lash on some engines is adjusted with an adjusting nut on the valve tip end of the rocker arm. The clearance is checked by inserting a feeler gauge between the valve tip and the adjusting nut. Some overhead cam (OHC) engines have an adjustment disc or shim between the cam lobe surface and the lifter or follower. To adjust valve lash, a special tool and a magnet must be used.

Prior to checking or adjusting valve clearance, the technician must consult the vehicle service manual to determine the type of valve train adjustment method used and whether or not the adjustment must be done with the engine cold or warm. Many late-model engines use hydraulic lash adjusters that eliminate the need for periodic adjustment. If the engine requires periodic valve adjustment, then the exact procedures must be followed to prevent incorrect adjustment and possible valve train damage. If valves are adjusted with insufficient clearance, the valve may hold open when the engine is warm, and this will cause a loss of compression and lead to valve leakage and burning. If the valve clearance is excessive, then the valve will open late and not open as far as it should. This condition causes the valve timing to be retarded, and valve overlap and lift are reduced. This will result in lower power contribution from the cylinder.

12. Verify camshaft timing; verify operation of camshaft timing components, including engines equipped with variable valve timing; determine needed action.

The camshaft and crankshaft must always remain in the same relative position to each other. They must also be in the proper initial relation to each other. This initial position between the shafts is designated by timing marks. To obtain the correct initial relationship of the components, the corresponding marks are aligned during engine assembly. Verification of this relationship is done by rotating the crankshaft to TDC on cylinder #1 and checking the alignment of the timing marks on both shafts.

Problems such as a no-start condition, lack of engine power, or a low vacuum reading on a vacuum gauge should alert the technician to verify the correct valve timing relationship. On distributor-equipped engines on which the distributor is driven by the camshaft, if the ignition timing is found to be off by more than a couple of degrees, the camshaft timing should be checked to make sure the timing chain or belt has not jumped or been installed incorrectly. To verify correct valve timing on vehicles with a distributor and chain-driven camshaft, the engine should be rotated until the ignition timing marks are aligned at

TDC. The distributor cap should then be removed and the ignition rotor should be pointing to spark plug #1 wire terminal.

On engines with a belt-driven camshaft, the timing covers are usually removed to inspect the timing marks on the camshaft and crankshaft, or rubber access plugs may be present to allow inspection. On distributorless ignition engines, the valve cover may have to be removed to inspect valve operation and position for cylinder #1 when the engine is rotated to TDC on cylinder #1 compression stroke.

13. Diagnose emissions or driveability problems caused by oil related issues, such as incorrect pressure, poor quality, incorrect level, or incorrect type used for the application.

Engine oil is the lifeblood of any engine. Failure to change it at regular maintenance intervals and at the correct level and viscosity can cause problems with the oil-controlled variable valve timing system, which in return will cause emission and/or driveability concerns. Along with controlling the variable valve timing system on some vehicles, engine oil lubricates all moving parts in the engine, helps cool the engine, reduces friction, cleans and holds dirt, and helps seal the piston rings. Oil that becomes dirty and sludgy can cause the cam phaser and oil control valves to stick, preventing the intake and exhaust valves from operating at different points in the combustion cycle to improve engine performance.

14. Verify engine operating temperature, check coolant level and condition, perform cooling system pressure test; determine needed repairs.

The cooling system must operate, be inspected, and be serviced as a system. Replacing one damaged part while leaving others dirty or clogged will not increase system efficiency. Service the entire system to ensure good results. Service involves both pressure testing and a visual inspection of the parts and connections. Pressure testing is used to detect internal or external leaks. Pressure testing can also be used to check the condition of the radiator cap. This type of testing involves applying a pressure to the system or cap. If the system is able to hold the pressure, there are no leaks in the system. If the pressure drops, then there is an internal or external leak.

The vehicle's temperature gauge, a shop temperature gauge, or a hand-held pyrometer can verify engine operating temperature. The condition and level of the coolant should be checked as part of a preventative maintenance program. The level of the coolant should be at the level specified by the manufacturer. The coolant should be checked for the presence of engine oil or other contaminants.

15. Inspect and test mechanically/hydraulically/electronically operated fans, fan clutch, fan shroud/ducting, and fan control devices; determine needed repairs.

Mechanical fans can be checked by spinning the fan by hand. A noticeable wobble or any blade that is not in the same plane as the rest indicates that the fan needs to be replaced. One of the simplest checks of a fan clutch is to look for signs of fluid loss. Oily streaks radiating outward from the hub shaft mean fluid has leaked out past the bearing seal. Most fan clutches offer a slight amount of resistance if turned by hand when the engine is cold. They offer drag when the engine is hot. If the fan freewheels easily when it is hot or cold, then replace the clutch.

Electric cooling fans are mounted to the radiator shroud and are not mechanically connected to the engine. An electric motor-driven fan is controlled by an engine coolant temperature switch or sensor, an air conditioning switch, or both. The controls for the electric cooling fan can be easily identified by referring to the vehicle's wiring schematic.

16. Read and interpret electrical schematic diagrams and symbols.

Electrical schematics, also known as electrical diagrams or wiring diagrams, are the roadmaps technicians use to properly diagnose and understand any electrical circuit on the vehicle. By studying and completely understanding an electrical schematic, many technicians are able to identify electrical problems with no other diagnostic information. Understanding the symbols used on the schematics is a major key to fully utilizing the diagrams as a diagnostic tool. A technician must be able to identify where the circuit power originates, whether the device is voltage-controlled or ground-controlled, and what other devices are powered from the same source. Identifying terminal pin numbers, wire colors, which fuse(s) feed the circuit and their location, circuit identifying numbers, splices, connectors, and where a circuit is grounded are all quite necessary in nearly every type of electrical operation. All are possible if a technician has access to the proper wiring schematic.

Most diagnostic procedures are written with the assumption that the technician performing the test is familiar with the location and types of circuits being tested. Having a specific electrical schematic of that circuit in front of the technician, and his ability to gain information from the schematic will ensure that the technician is able to identify what is being tested with the most efficiency.

Many computer or Internet-based information systems provide the ability to print out complete schematics or even to zoom in on a specific portion of it and print the zoomed image. The technician can then write notes on his own printed copy or even highlight current flow or other pertinent items that he wishes to focus on.

17. Test and diagnose emissions or driveability problems caused by battery condition, connections, or excessive key-off battery drain; determine needed repairs.

Battery voltage is not only needed to start the engine, but is also very important in stabilizing the voltage during engine operation. Low battery voltage or state of charge, as well as poor battery cable connections, will cause slow engine cranking and hard- or no-start complaints. Once the engine is running, the vehicle's charging system supplies the voltage needed and restores the charge of the battery. As the charge voltage is supplied to the system, it will fluctuate depending on vehicle electrical loads and the sensed need of the battery. Once the battery has the same voltage as the output of the charging system, the system voltage and charging output is leveled. Many systems on today's vehicles are monitored by electronic components that send information to a control module or computer. This information is delivered as changes in voltage. Therefore, precise voltage control is important to effective engine control management. When there is a large fluctuation in system voltage of voltage spikes, the computer may reset base sensor input levels stored in memory, and drivability problems may result.

Most computers have a few milliamperes of current draw when they are not in operation. This current draw is called parasitic load. Since many vehicles today have several computers, this current draw may discharge a battery if the vehicle is not driven for several weeks. Most manufacturers allow between 50 and 100 milliamps parasitic current draw. A value greater than 100 milliamps is considered excessive and further diagnostics should be performed to determine what is causing the current drain.

Connecting a digital multimeter (DMM) set to measure amperage in series between the negative battery cable and battery post will allow the technician to measure this load. All accessories must be off and the key removed from the ignition. Some vehicles may require a long wait period of up to one hour before all computers will power down. Take a reading and compare it to manufacturer's specifications.

18. Perform starter current draw test; determine needed action.

Starter current draw testing is only performed on batteries with a *specific gravity* of 1.190 or greater. Several different testers can be used. If an analog tester is used, always check the mechanical zero on each meter and adjust as necessary. Be sure that all electrical loads are off and the doors are closed, as additional loads will cause additional draw. The ignition is disabled and the engine is cranked while observing the ammeter and voltmeter readings. High current draw and low cranking speed usually indicate a defective starter. High current draw may also be caused by internal engine problems. A low cranking speed and low current draw with high cranking voltage usually indicate excessive resistance in the starter circuit, such as in the cables and connections.

19. Perform starter and charging circuit voltage drop tests; determine needed action.

The resistance in an electrical wire may be checked by measuring the voltage drop across the wire with normal current flow in the wire. To measure the voltage drop across the positive battery cable, connect the positive voltmeter lead to the positive cable at the battery, and connect the negative voltmeter lead to the other end of the positive battery cable at the starter solenoid. Disable the ignition system. Crank the engine. The voltage drop indicated on the meter should not exceed 0.5 volts. If the voltage reading is higher than this figure, then the cable has excessive resistance. If the cable ends are clean and tight, then replace the cable. Connect the positive voltmeter lead to the positive battery cable on the starter solenoid, and connect the negative voltmeter lead to the starting motor terminal on the other side of the solenoid. Leave the voltmeter on the lowest scale and crank the engine.

If the voltage drop exceeds 0.3 volts, then the solenoid disc and terminals have excessive resistance.

Connect the positive voltmeter lead to the starter motor housing and the negative lead to the negative battery post and crank the engine. If the reading is greater than 0.2 volts and the cable connections are clean and tight, then replace the negative battery cable.

20. Test and diagnose engine performance problems resulting from charging system failures; determine needed action.

The charging system is responsible for maintaining stable electrical system voltage. Undercharging as well as overcharging may cause engine performance problems including hard starting or battery gassing and corrosion concerns. A defective diode in the alternator may allow enough AC voltage leakage into the electrical system to disrupt normal computer operation. Test the alternator with a DMM or lab scope for excessive AC voltage output; replace it if it is found to be out of specifications.

If alternator output is zero, then the alternator field circuit may be open. The most likely place for an open circuit in the alternator is at the slip rings and brushes. If alternator output is normal, but no charging current is measured at the battery, then the fuse link between the alternator output terminal and positive battery cable is probably open. If the alternator output is less than specified, always be sure that the belt and belt tensions are satisfactory. If belt condition and tension are satisfactory and the alternator output is less than specified, then the alternator is defective. On some alternators, there is a method of "full fielding" the unit. This technique will bypass the voltage regulator circuit and full alternator output will be obtained. In this case, if the alternator has full output, then the regulator or its circuit has failed.

21. Inspect, adjust, and replace alternator (generator) drive belts, pulleys, clutches, tensioners, and fans.

A loose belt causes low alternator output and a discharged battery. A loose, dry, or worn belt may cause squealing and chirping noises during acceleration and cornering. Belt tension may be checked by measuring the belt deflection. Press on the belt with the engine stopped to measure the belt deflection. A free span of .5 in (12.7 mm) per foot (30.5 cm) is usually acceptable.

Serpentine drive belt systems will most often use automatic belt tensioners. Automatic tensioners are usually spring loaded, but should be checked to be sure they are applying adequate pressure to the belt to prevent slipping or squealing from the belt. Pulley alignment is critical on serpentine belt drives. Any misalignment will cause noise from the belt and may allow the belt to slip off the pulleys.

Whenever a customer with a serpentine drive belt engine consistently complains of belt noise, the technician should check to see if any previous service has been performed on the engine that may have required accessories to be removed. If a pulley is reinstalled backward or any spacers are left off of accessory brackets, then pulley misalignment can occur and cause constant belt noise. Proper routing of the serpentine belt must be observed when replacing a drive belt. Routing the belt incorrectly can cause some accessories to spin backward or not allow proper tension to be applied to the belt by the belt tensioner.

22. Inspect, test, and repair or replace charging circuit components, connectors, and wires in the starter and charging control circuits. Inspect, test, and repair or replace components and wires in the starter control circuit.

Check wires for burned or melted conditions. Check connector ring terminals for loose retaining nuts, which cause high resistance or intermittent open circuits. An open circuit may be caused by a terminal that is backed-out of the connector. Terminals that are bent or damaged may cause shorts or open circuits. An open circuit occurs when the terminal is crimped over the insulation instead of the wire core. A greenish white corrosion on terminals results in high resistance or an open circuit.

When testing the starter control circuit, connect the positive voltmeter lead to the positive battery cable at the battery, and connect the negative voltmeter lead to the solenoid winding terminal on the solenoid. Leave the ignition systems disabled and place the voltmeter selector on the lowest scale. If the voltage drop across the control circuit exceeds 1.5 volts while cranking the engine, individual voltage drop test on control circuit components are necessary to locate the high-resistance problem.

B. Ignition System Diagnosis and Repair (8 Questions)

1. Diagnose ignition system related problems such as no starting, hard starting, engine misfire, poor driveability, spark knock, power loss, poor mileage, and emissions problems; determine root cause; determine needed repairs.

Diagnosing engine performance and no-start problems related to the ignition system requires a thorough understanding of the type of ignition system used and the components used in the primary and secondary circuits. Engines will use either a *distributor ignition* (DI), electronic distributorless ignition using a coil for each pair of spark plugs and called "waste spark" systems electronic distributorless ignition (EI), or a *coil on plug* (COP) ignition system. Most ignition systems used since the mid-eighties will incorporate computer control of ignition timing.

No-start diagnosis involves checking for the presence of spark with a spark tester first. If no spark is present at the plugs, then the problem must be isolated to either the primary or secondary circuit. Testing the primary circuit with a test light or scope will determine if the coil is being switched on and off. If the coil does not switch on and off, then the problem is in the primary circuit and a bad triggering device, module, or wiring is the cause. If the coil is being switched on and off, then the coil secondary windings, a distributor cap and rotor, or secondary wiring is at fault.

Further pinpoint testing of ignition system performance can be performed with a scope. Testing for available secondary voltage and performing a wet test by spraying water on the secondary wiring will help isolate problems that can cause many different driveability concerns like misfire and hard starting. Consulting specific manufacturer's test procedures will be necessary when troubleshooting concerns such as spark knock, poor mileage, or emissions problems due to the many different computer-controlled systems in use.

Defects in the ignition system or incorrect adjustments can cause many common driveability problems. Overadvanced ignition timing will cause pinging or spark knock and increase hydrocarbon emissions, while retarded ignition timing will cause a lack of power and poor mileage. Worn or damaged secondary ignition components can cause engine misfire or rough running complaints and will increase tailpipe emissions. Secondary voltage leakage from damaged insulation on spark plug wires or cracked spark plug insulators can cause bucking and hesitation complaints that may only occur under road load conditions.

When diagnosing ignition system problems, the technician must always consider what caused a failure to occur. For instance, when an ignition module fails, it may be due to increased current flow from a shorted primary winding in the ignition coil. Replacing the ignition module may allow the car to start, but the replacement module will fail prematurely. By performing a more thorough testing routine and checking the ignition coil windings for proper resistance, the root cause of the module failure would be uncovered. Often when dealing with electronic component failures, there is another component problem that will cause a related part to fail. Finding and correcting the root cause of a failure is the difference between a parts changer and a technician.

2. Interpret ignition system-related diagnostic trouble codes (DTCs); determine needed repairs.

Diagnostic trouble codes (DTCs) can be retrieved from the powertrain control module (PCM) on most vehicles. The DTCs are displayed in the instrument panel or on a scan tool. The latter is most common. Displayed trouble codes are identified in the service manual along with test procedures necessary to isolate the cause. A DTC usually identifies a problem area, not the exact cause. Performing the pinpoint tests found in the service manual should determine the exact problem causing the code to set. Some vehicles have more in-depth ignition system trouble code capabilities and can set DTCs for primary ignition components like modules or triggering devices while other systems may only set codes for spark timing control problems.

3. Inspect, test, repair, or replace ignition primary circuit wiring and components.

Ignition system primary wiring faults will most often be seen as bad connections at terminals and connectors. Probing wiring through insulation to make electrical tests will damage the insulation and can lead to wiring corrosion and high resistance, especially in climates where road salt is used in the winter. The technician must test primary wiring for unwanted resistance and current carrying capacity by performing available voltage and voltage drop tests or measuring resistance with an ohmmeter. When replacing primary wiring, the correct gauge wire must be used and soldering connections and sealing the repair with heat shrink tubing should provide adequate weatherproofing. Testing primary components such as the ignition switch or ballast resistor, if used, can be done with a DMM. Both power and ground circuits must be checked to insure proper primary circuit current flow and operation.

4. Inspect, test, service, repair, or replace ignition system secondary circuit wiring and components.

Ignition system secondary circuit wiring should be visibly inspected for chafing against metal brackets or exhaust manifolds that could cause arcing as well as proper routing. Misrouted wires can increase the possibility of cylinder crossfire that can cause extreme engine damage. Wires and spark plug boots should also be checked for coolant or oil soaked conditions and replaced if these conditions exist. Spark plug wire terminals should also be checked for signs of corrosion or arcing at both ends of the wire. Wires can be checked for proper resistance with an ohmmeter; specifications are listed in the service manual and are usually expressed in ohms per foot.

Distributor caps and rotors should be inspected for burned terminals and cracks as well as secondary voltage arcing damage. An ignition oscilloscope can be used to check for leakage in secondary wiring. Spraying a fine mist of water on the wires to simulate damp conditions that cause many secondary ignition problems should be done. Worn spark plug gaps or excessive distributor cap to rotor air gaps are easily tested with an ignition scope.

5. Inspect, test, and replace ignition coil(s).

The ignition coil should be inspected for cracks or any evidence of leakage in the coil tower. The coil container should be checked for oil leaks. If oil is leaking from the coil, then air space is present, allowing condensation to form internally. Condensation in an ignition coil causes voltage leaks and engine misfire. When testing the coil with an ohmmeter, most primary windings have a resistance of 0.5 ohms to 2 ohms and secondary windings have a resistance of 8,000 ohms to 20,000 ohms. The maximum coil output can be tested with an engine analyzer. Always refer to the manufacturer's specifications.

6. Inspect, test, and replace ignition system sensors; adjust as necessary or triggering devices.

There are several different styles of primary triggering devices used on ignition systems. Magnetic pulse generators, Hall-effect switches, and optical pickup are the most common. When testing magnetic pickups an ohmmeter can be used to test for proper resistance of the coil and for a grounded or open coil. Connect the ohmmeter leads across the coil leads to check for proper resistance or an open coil. Connect one meter lead to a pickup coil lead and the other to ground to test for a grounded coil. Other tests that can be performed on magnetic pickups include testing AC voltage output with a DMM and signal waveform measurements with a scope. Consult the service manual for specifications for these tests.

Hall-effect switches and optical pickups require a voltage feed and ground to operate properly. Once proper feed voltage and a good ground are verified, the signal line can be tested. These pickups are often supplied a reference voltage from the ignition control module or powertrain control computer, usually between 5 and 10 volts. A lab scope is the best method for testing the signal from these types of pickups.

Some Hall-effect and optical pickups may have four wires. In this case, there will be two signal lines on the lab scope trace; often these will be low and high data rate signals. Some optical pickups generate a square wave signal output for every degree of camshaft rotation, so the signal created will be a very high frequency. While a DMM that measures frequency can determine if a signal is being generated, a lab scope provides a much better picture of signal integrity. The square wave signal from a Hall-effect or optical pickup should reach 90 percent of reference voltage when the signal is high and pull down to within 10 percent of ground potential when the signal is low.

7. Inspect, test, and/or replace ignition control module (ICM)/power train/engine control module (PCM/ECM); reprogram as needed.

Many ignition module testers are available from vehicle and test equipment manufacturers. These testers check the module's capability of switching the primary ignition circuit on and off. On some testers, a green light is illuminated if the module is satisfactory, and the light remains off when the module is defective. Always follow the manufacturer's recommended procedure. If the module tests satisfactory, then the technician should perform circuit tests to confirm that the wiring of the circuit is serviceable and the proper signals are going to the correct designations.

The ignition module removal and replacement procedure varies, depending on the ignition system. Always follow the manufacturer's recommended replacement procedure. Some ignition modules require the use of dielectric silicone grease for heat dissipation through the mounting surface. Clean the mounting surface and apply a light coat of silicone grease to the module prior to installing the module. If silicone is not used, then heat generated by the modules transistor will not be properly dissipated, and early failure of the module may occur.

Replacement of a powertrain control module should be done only after power and ground circuits to the PCM have been tested and all PCM outputs have been checked for proper current draw to prevent repeated failure. The technician must be careful to ground himself or herself to dissipate any static electrical charges from their body prior to handling the replacement PCM. This will prevent damage to sensitive circuit boards from any static electrical discharge.

C. Fuel, Air Induction, and Exhaust System Diagnosis and Repair (9 Questions)

> *Note:* Fuel and air induction system diagnosis and repair will include vehicles equipped with speed density (MAP) and mass airflow (MAF) systems. These applications will be identified in ASE tests.

1. Diagnose fuel system-related problems, including hot or cold no-starting, hard starting, poor driveability, incorrect idle speed, poor idle, flooding, hesitation, surging, engine misfire, power loss, stalling, poor mileage, dieseling, and emissions problems; determine root cause; determine needed action.

The first step in diagnosing fuel system-related problems is to identify the type of system used on the vehicle and review its theory of operation. A clear description of the problem must be obtained, and the technician should attempt to verify the complaint. After the problem is verified, the technician should check for any relative service bulletins before proceeding with detailed testing. Once the complaint is isolated to the fuel system, the technician can perform pinpoint tests of the fuel delivery system to determine the exact cause of the problem.

The service technician must always seek to determine the root cause of a failure to prevent repeated problems. An example is a fuel pump that fails due to debris in the fuel tank. Replacing the pump may correct the driveability problem, but the root cause of the fuel pump failure was dirt in the fuel tank. If the contamination is not found and removed from the tank, the new fuel pump will fail prematurely.

Problems related to a lack of fuel delivery include power loss, hesitation, surging, and hard starting. Excessive fuel delivery from items such as high fuel pressure, leaking injectors, or a leaking or stuck fuel pressure regulator can cause a no- or hard-start condition, flooding, engine misfire and poor idle quality, poor fuel economy, and excessive exhaust emissions. Any restrictions in the air inlet or exhaust system can cause reduced engine power, poor mileage, and surging or hesitation complaints.

2. Interpret fuel or induction system-related diagnostic trouble codes (DTCs); analyze fuel trim and other scan tool data; determine needed repairs.

Fuel system-related diagnostic trouble codes can be retrieved from nearly all computer-controlled vehicles. Older systems support both flash code and scan tool diagnostics, while 1996 and newer vehicles use primarily scan tool code retrieval. Fuel system trouble codes rarely point to specific components, but most often are set because the computer can no longer properly control the air/fuel mixture at ideal levels. Once trouble codes are obtained, the service manual should be consulted to perform the tests necessary to pinpoint the exact cause of the problem.

Often diagnostic trouble codes point to a system, not a component, and the technician must consider what the computer sees in order to set a trouble code. An example would be a lean exhaust trouble code. If the oxygen sensor fails, then it no longer produces an output voltage. The computer sees the low signal voltage and interprets this as a lean exhaust signal. The computer will increase fuel delivery in an attempt to generate a high voltage signal from the oxygen sensor. The engine can be running very rich and an exhaust gas analysis will confirm this, yet the computer trouble code is set for a lean exhaust signal because of the failed oxygen sensor.

3. Inspect fuel tank, filler neck, and gas cap; inspect and replace fuel lines, fittings, and hoses; check fuel for contaminants and quality; determine needed repairs.

The fuel tank should be inspected for leaks; road damage; corrosion; rust; loose, damaged, or defective seams; loose mounting bolts; and damaged mounting straps. Leaks in the fuel tank, lines, or filter may cause gasoline odor in and around the vehicle, especially during low speed driving and idling. In most cases, the fuel tank must be removed for service.

Nylon fuel pipes should be inspected for leaks, nicks, scratches, cuts, kinks, melting, and loose fittings. If these fuel pipes are kinked or damaged in any way, then they must be replaced. Nylon fuel pipes provide a certain amount of flexibility and can be formed around gradual curves under the vehicle. Do not force a nylon fuel pipe into a sharp bend, because this action may kink the pipe and restrict the flow of fuel.

Obtain a sample of the fuel and examine for dirt or other contaminants. Use a commercially available alcohol tester to determine the percentage of alcohol present in the fuel. Generally fuel injection systems will tolerate only a limited amount of alcohol in the fuel. Consult workshop manual for details.

4. Inspect, test, and replace fuel pump(s) and/or fuel pump assembly; inspect, service, and replace fuel filters.

When testing mechanical and/or electric fuel pumps, pressure and volume tests apply. Fuel must be available to the engine at the correct pressure and adequate volume for proper operation. It is possible to have the correct pressure with little or no volume. Fuel volume is the amount of fuel delivered over a specified period of time. The correct amount is specified by the manufacturer and is generally about 1 pint in 30 seconds. Caution should be exercised when performing volume testing, because fuel is discharged into an open container, which creates a risk of fire.

Mechanical fuel pumps can be tested for both vacuum and pressure by alternately connecting a vacuum/pressure gauge first to the inlet fitting with the outlet removed and cranking the engine, then reconnecting the inlet line and starting the engine with the gauge teed into the outlet line to the carburetor. Electric fuel pumps can be tested for proper pressure by connecting a fuel pressure gauge to the Schrader valve test port on the fuel rail. If the vehicle does not have a test port on the fuel rail, then the gauge will need to be teed into the system using the correct adaptors.

Low pressure can result from a restricted fuel line or filter, a defective pump, or a lack of good voltage supply and ground connections in the pumps electrical circuit. If fuel pressure is higher than specified, then a bad fuel pressure regulator or a restricted fuel return line could be the cause. Most fuel injection systems should hold pressure after the engine is turned off. A bad fuel pump outlet check valve, a bad fuel pressure regulator, or a leaking fuel injector can cause a pressure leak down. Follow the manufacturer's test procedures to isolate the failed component.

Fuel filters should be replaced at the manufacturer's specified intervals—more often if the fuel quality in the area is poor or if the vehicle has had fuel contamination problems.

5. Inspect and test electric fuel pump electrical control circuits and components; determine needed repairs.

Electric fuel pumps are supplied voltage through a variety of different control circuits and components depending upon the manufacturer. Most commonly, the computer will control a relay that supplies current to the pump. The computer will energize this relay when the key is first turned on for two seconds to prime the system. If the engine is not started, then the computer shuts off the relay after the prime period. If the engine were to stall, then the computer will again turn off the relay. Once the computer receives an RPM signal indicating the engine is running, the fuel pump relay will remain energized. Many cars also use an inertia switch in the feed line to the pump. In case of an accident, the inertia switch will open the circuit preventing fuel pump operation and reducing the risk of a fire.

On some import vehicles, the fuel pump control circuit goes through contacts in the vane-style airflow meter. Once again with this system, if the engine does not stay, then running the fuel pump will not operate.

If there is a fuel delivery problem or no-start condition because of a lack of fuel, then the technician must obtain wiring schematics and determine the type of control circuit used to power the fuel pump. With this information, the technician can perform the necessary electrical tests on the components in the system to isolate the problem.

6. Inspect, test, and repair or replace fuel pressure regulation system and components of fuel injection systems; perform fuel pressure/volume test.

The fuel pressure regulator should be tested for leakage through the diaphragm by removing the vacuum hose with the engine running to see if fuel drips from the vacuum nipple. Fuel pressure should reach the pressure specified by the manufacturer with the vacuum line disconnected and lower by about 10 psi when the vacuum line is plugged back onto the regulator. Excessive fuel pressure indicates a stuck closed regulator or restricted fuel return line. Low pressure can be caused by a weak fuel pump, restricted fuel filter, or stuck-open regulator. If the pressure increases to normal when the return line is restricted, then the regulator should be replaced.

If fuel pressure drops when the engine is turned off, then a leak is indicated in the fuel system. Alternately clamping off the fuel feed and return lines will isolate the location of the leak. If the pressure drop stops when the return line is pinched, then the fuel pressure regulator is leaking and should be replaced. If the pressure still drops, then a leak through the fuel pump or injectors is the problem. Clamping the fuel feed line will eliminate the fuel pump as the source of the pressure drop if the system holds pressure; if that is the case, then the injectors will need service. Do not clamp a plastic fuel line, as this will cause permanent damage. Rubber test lines need to be installed to perform this test if the vehicle has plastic fuel lines.

7. Inspect, remove, service or replace throttle assembly; make related adjustments and/or perform initialization or relearn procedure as required.

After many miles of operation, an accumulation of gum and carbon deposits may occur around the throttle area on throttle body-injected (TBI), multiport fuel-injected (MFI), and sequential fuel-injected (SFI) systems. This can cause rough idle and stalling problems. Throttle body cleaner may be used to spray around the throttle area without removing and disassembling the throttle body. If this cleaning does not remove the deposits, then the throttle body will have to be removed according to the manufacturer's recommendations, disassembled, and placed in an approved cleaning solution. The throttle position sensor (TPS), idle air control (IAC), fuel injector, pressure regulator, and seals must be removed prior to placing the throttle body into the cleaning solution. Since MFI and SFI systems do not have the pressure regulator or fuel injectors in the throttle body, these items need not be removed. After cleaning, the closed throttle position or minimum air rate will need to be checked and adjusted, if necessary, along with setting the throttle position sensor voltage. Consult the service manual for the correct procedures to follow when making these adjustments.

8. Inspect, test, clean, and replace fuel injectors and fuel rails.

Fuel injectors can be tested on the car by performing an injector balance test. While monitoring fuel pressure, each injector is fired one at a time with a special tool designed to open the injector an exact amount of time. When the injector is triggered, the fuel pressure in the rail will drop. No pressure drop indicates a plugged injector or open coil. An injector with a low-pressure drop indicates a dirty or restricted injector. An injector with too great a pressure drop indicates a leaking or rich injector. A pressure difference greater than 1.5 psi above or below the average is considered a problem that requires service. Fuel injectors should also be tested with a lab scope to observe their electrical integrity. Both voltage and current waveforms can be observed with a lab scope.

Tool manufacturers market a variety of fuel injector cleaning equipment. A solution of fuel injector cleaner is mixed with unleaded gasoline. Shop air pressure provides system operating pressure. The vehicle's fuel pump must be disabled to prevent fuel from being forced into the fuel rail. The fuel return line should be plugged to keep the fuel injector cleaner solution from entering the fuel tank. After the fuel injectors have been cleaned, the adaptive memory will need to be reset.

If the injectors have been removed for cleaning, then the spray pattern should be checked. An even cone-shaped pattern without thready or dripping discharge should be present. If the proper spray pattern cannot be achieved, the injector should be replaced.

9. Inspect, service, and repair or replace air filtration system components.

If a vehicle is operated continually in dusty conditions, then air filter replacement may be necessary at more frequent intervals. A damaged air filter can cause increased wear on cylinder walls, pistons, and piston rings. When the air filter is restricted with dirt, it restricts the flow of air into the intake manifold, and this increases fuel consumption.

10. Inspect throttle assembly, air induction system, intake manifold and gaskets for air/vacuum leaks, restrictions, and/or unmetered air.

Intake manifold vacuum leaks or restrictions can cause rough idle and stalling complaints or incorrect idle speeds, and may cause trouble codes to be logged for idle speed or fuel trim errors. Vacuum leaks can be located by flowing propane around suspected areas to see if the idle is affected. A smoke machine is a popular tool for locating vacuum leaks and can be used without running the engine. Not only will intake leaks be located with a smoke machine, but any vacuum accessory connected to the intake manifold will also be checked and pinpointed if leaking. A careful visual inspection should be performed on all air induction hoses to locate any cracks, restrictions, or loose clamps. This is especially important on mass airflow-equipped engines, because unmetered air drawn into the engine behind the airflow meter will cause lean running conditions and a lack of power or hesitation complaints.

11. Remove, clean, inspect, test, and repair or replace fuel system vacuum and electrical components and connections.

Fuel system vacuum and electrical components include the fuel pressure regulator and any vacuum controls if used, vacuum-operated throttle positioner, fuel pump relay, inertia switch, two-speed fuel pump resistor, and electronic fuel pump power modules. A visual inspection will uncover damaged vacuum lines, and proper routing can be checked against the under-hood emissions label. All electrical connections should be visually checked for terminal seating as well as damaged, chafed, or corroded wires or connections. Basic electrical testing with a test light and DMM can determine problems with fuel system electrical components. The service manual should be consulted to identify which components are used and what, if any, special test procedures may be required.

12. Inspect, service, and replace exhaust manifold, exhaust pipes, oxygen sensors, mufflers, resonators, catalytic converters, resonators, tail pipes, and heat shields.

Remove the exhaust pipe bolts at the manifold flange, and disconnect any other components in the manifold, such as the O_2 sensor. Remove the bolts retaining the manifold to the cylinder head, and lift the manifold from the engine compartment. Remove the manifold heat shield. Thoroughly clean the manifold and cylinder head mating surfaces. Measure the exhaust manifold surface for warping with a straightedge and feeler gauge in three locations on the manifold surface. Examine the manifold carefully for any cracks or broken flanges.

Follow the exhaust system from manifold to tail pipe end. Ensure that all hangers are present and installed correctly. The exhaust system is designed to be suspended from these hangers; loosen joints and realign if any of the hangers are in tension. Examine all pipes, mufflers, and resonators to ensure that they are securely connected and gas tight.

13. Test for exhaust system restriction or leaks; determine needed action.

A restricted exhaust pipe, catalytic converter, or muffler may cause excessive exhaust back pressure. If the exhaust back pressure is excessive, then engine power and maximum vehicle speed are reduced. Even a partial restriction will reduce performance and fuel mileage. A low-range pressure gauge or compound vacuum gauge can be connected into the exhaust system to measure back pressure directly. Back pressure test adaptors are available that screw into the oxygen sensor hole and provide a hose nipple for connection of a compound vacuum gauge. Another adaptor looks like a rivet that is installed into a hole drilled into the exhaust system front pipe. This adaptor allows a gauge to be screwed into the inside threads of the adaptor. Generally, exhaust back pressure should measure less than 3 psi maximum when engine speed is held around 2,500 rpm.

Another method of testing for exhaust restrictions is to measure intake manifold vacuum. Normal vacuum at idle should be between 16 and 21 in. hg (48.3 to 31 kPa absolute). When the engine is accelerated to 2,500 rpm and held, the vacuum reading will drop momentarily and then stabilize equal to or greater than the idle reading. A vacuum reading that drops very low or to zero indicates a restricted exhaust system.

14. Inspect, test, clean, and repair or replace turbocharger or supercharger and system components.

Both a *turbocharger* and *supercharger* increase the amount of air delivered to an engine's cylinders by increasing the amount of pressure at which the air is delivered. A turbocharger is driven by the velocity and heat of the exhaust leaving the engine. Intake air pressure is increased by a compressor in the turbocharger unit. The faster the compressor turns, the more the air is boosted. The speed of the compressor is determined by the load and speed on the engine. To control the boost and therefore prevent over-boost, turbochargers are equipped with a wastegate that controls the amount of exhaust gas at the turbocharger.

A supercharger is driven by the engine's crankshaft via a drive belt. The speed of the compressor or supercharger is directly related to the speed of the engine.

The pressure boost for either system can be measured with a pressure gauge connected to the intake manifold. During a road test, the pressure can be observed during a variety of speed and load conditions. Recording the condition and the resulting pressure can lead to a thorough evaluation of the turbocharger or supercharger system.

Both a supercharger and turbocharger are non-serviceable items. If there is a problem with either unit, then it is replaced. Only the control circuits of these systems can be serviced.

D. Emissions Control Systems Diagnosis and Repair (including OBD II) (8 Questions)

1. Positive Crankcase Ventilation (PCV) (1 Question)

1. Test and diagnose emissions or driveability problems caused by PCV system.

If the positive crankcase ventilation (PCV) valve is stuck in the open position on a carbureted engine, then excessive airflow through the valve causes a lean air/fuel ratio and possible rough-idle operation or engine stalling. On a fuel-injected engine, a stuck open PCV valve can cause high idle speed complaints. When the PCV valve or hose is restricted, excessive crankcase pressure forces engine blow-by gases through the clean air hose and filter into the air cleaner housing.

Oil accumulation and crankcase blow-by gases may also be found in the air cleaner housing on high mileage engines, because excessive engine blow-by pressure from worn piston rings and cylinders will create more crankcase pressure than the PCV system can handle. Internal engine repairs are the only means of correcting this problem. Flow-meter-style testers are available to test for excessive crankcase blow-by.

2. Inspect, service, and replace PCV filter/breather cap, valve, oil separator tubes, orifices/metering device, and hoses.

A thorough examination of the PVC system is relatively easy. After performing the recommended diagnostics, visually inspect the cap, tubes, and hoses for kinks, cuts, or other damage. Disassemble the PVC system to isolate the cause of the restriction. Shake the PCV valve next to your ear and listen for the tapered valve rattling inside the housing. If no rattle is heard, replace the PCV valve.

PCV diagnostic recommendations differ from manufacturer to manufacturer. Some recommend removing the PCV valve and hose from the rocker cover. Connect a length of hose to the inlet side of the PCV valve, and blow air through the valve with your mouth while holding your finger near the valve outlet. Air should pass freely through the valve. If not, then replace the valve. Connect a length of hose to the outlet side of the PCV valve and try to blow back through the valve. If air passes easily through the valve, then it should be replaced. Other manufacturers recommend disconnecting one end of the PCV valve and placing a finger over it with the engine idling. When there is no vacuum at the PCV valve, part of the system is restricted.

2. Exhaust Gas Recirculation (2 Questions)

1. Test and diagnose driveability problems caused by the exhaust gas recirculation (EGR) system.

The EGR valve should open once the engine is warm and run above idle or under road load conditions. If the EGR valve remains open at idle and low engine speed, then rough idle and stalling can occur as well as engine surging during low-speed driving conditions. When this problem is present, the engine may also hesitate on low-speed acceleration or stall after deceleration or after a cold start. If the EGR valve does not open, then engine detonation can occur and emissions levels will increase. EGR system problems can affect emissions levels differently, depending on what the problem is. A stuck open EGR valve will create a density misfire condition in which the air/fuel mixture is diluted with exhaust gas, causing misfire in the combustion chamber. This problem will increase hydrocarbon emissions and raise oxygen levels. An EGR valve that does not open will cause an increase in oxides of nitrogen emissions.

By introducing exhaust gas into the engine during acceleration and cruise conditions, the inert exhaust gas helps reduce peak cylinder combustion temperatures below 2,500°F. When temperatures rise above 2,500°F, oxygen combines with nitrogen to form NOx, a main contributor to photochemical smog. It must be controlled to manageable levels inside the engine, because the three-way catalytic converter is not very efficient at reducing NOx. If the EGR valve does not open, then combustion chamber temperatures will rise and NOx production will increase far beyond what the catalytic converter can control.

2. Interpret EGR-related scan tool data and DTCs; determine needed repairs.

DTCs for the exhaust gas recirculation system will usually be set for one of three reasons: a control circuit fault, a no- or low-flow condition, or an excessive flow or flow when not commanded condition. The DTC should identify the problem area but not necessarily the component at fault. Once the DTC is retrieved, the service manual must be consulted to determine the necessary tests to be performed to isolate the cause of the code.

3. Inspect, test, valve service, and replace components of the EGR system, including EGR valve tubing, exhaust passages, vacuum/pressure controls, filters, hoses, electrical/electronic sensors, controls, solenoids, and wiring of EGR systems.

The first step in diagnosing any EGR system is to check all of the system's vacuum and electrical connectors. In many systems, the PCM uses inputs from various sensors to operate the EGR valve. Improper EGR operation may be caused by a defect in one or more of the sensors. DTCs should be retrieved and the cause corrected before any further diagnostics is completed.

There are both vacuum-operated and electronic EGR valves in use. Vacuum-operated EGR valves should be tested for proper vacuum supply to the valve. A vacuum pump should be used to apply vacuum to the valve to test the diaphragm and see that the valve opens. Some valves may require that the engine be running and off idle to operate the internal back pressure transducer so that the valve will hold vacuum. A noticeable change in engine speed and idle quality should be observed. This confirms the EGR passageways are not plugged. No change in engine operation will require removing the valve and cleaning the passages. A scan tool may be necessary to test electronic EGR valves properly. Vacuum bleed filters can be used on some systems and may require periodic replacement if they become clogged.

Often EGR problems are caused by faulty EGR controls, such as the EGR vacuum regulator (EVR). This regulator can be checked with a scan tool or an ohmmeter. Connect the meter across the terminals of the EVR. An infinite reading indicates there is an open inside the EVR, whereas a low resistance reading means the EVR's coil is shorted internally. The coil should also be checked for shorts to ground. To do this, connect the meter at one of the EVR terminals and the other to the case. The reading should be infinite. If there is any measured resistance, then the unit is shorted.

Other EGR control components can also be checked on the scan tool or with a DMM. Refer to the appropriate service manual for the exact procedures and the desired specifications.

3. Secondary Air Injection (AIR) and Catalytic Converter (2 Questions)

1. Test and diagnose emissions or driveability problems caused by the secondary air injection or catalytic converter systems.

Catalytic converters have been in use since 1975. They are placed in the exhaust system shortly after the engine and are designed to clean the exhaust of excess pollutants through a chemical reaction. There are three basic types of catalytic converters:

A. Two-way converters are used primarily on pre-1980 vehicles and control unburned hydrocarbons (HC) and carbon monoxide (CO).

B. Three-way converters (TWC) control HC, CO, and oxides of nitrogen (NOx). These converters are used on 1981 and later vehicles with computerized engine controls.

C. Three-way plus oxidation converters are used on 1980 and later vehicles with computerized engine controls and air injection.

The most common reasons for converter failure are overheating and contamination from oil burning or leaded fuel. An engine misfire or extremely rich air/fuel mixture can allow unburned fuel to enter the converter, which can cause excessive heat and converter failure.

Secondary air injection is used to reduce hydrocarbon (HC) and carbon monoxide (CO) emissions by oxidizing these pollutants in the exhaust manifold or catalytic converter. On some vehicles, outside air is injected into the exhaust manifold or converter by a belt-driven or electric air pump or by a pulse air-injection system. The air is routed through hoses and pipes by control valves and one-way check valves during certain engine operating conditions and mixed with exhaust gases as they leave the engine. Check valves installed at the exhaust manifold and catalytic converter air supply pipes prevent exhaust gases from flowing back into the control valves or air pump and damaging these components.

Air-injection control valves consist of a diverter valve (used to dump air pump output to atmosphere during deceleration to prevent backfiring) and switching valves that send air pump output to either the catalytic converter or exhaust manifold, depending on engine operating conditions.

On closed-loop feedback fuel control systems, the air pump output is directed to the exhaust manifold after starting and during warm-up. This allows faster warm-up of the oxygen sensor and catalytic converter. Once closed-loop fuel control is entered, the air pump output is switched to the catalytic converter to provide additional oxygen for the rear bed or oxidation bed of the converter. If a dual-bed converter is not used, then the air pump output is diverted to atmosphere. If the air pump fails or the hoses are disconnected, then tailpipe emissions will increase. If air is not diverted away from the exhaust during deceleration conditions, then exhaust backfiring may result.

Another type of secondary air injection is known as a *pulse air-injection system*. On this system, outside air is drawn into the exhaust manifold by negative pressure pulses created as the exhaust is pushed out of a cylinder by the piston. This system requires no power from the engine to run a pump, as with the belt-driven varieties.

A reed valve that is sensitive to the negative pressure pulses opens to allow airflow into the exhaust, but closes when positive exhaust pressure is present to prevent hot exhaust gases from backing up into the fresh air supply line that is usually connected to the air cleaner housing. This reed valve may also be known as an aspirator valve but the function is the same. Pulse air injection systems are most efficient at low engine speeds. At higher speeds, the exhaust pulses occur too rapidly, and very little air is drawn into the exhaust system.

2. Interpret secondary air injection system-related scan tool data and diagnostic trouble codes; determine needed repairs.

Some vehicles can set DTCs for secondary air injection system problems, such as control circuit fault codes, and airflow switching problems, such as air not being delivered to the exhaust manifold when commanded. Other vehicles may have no computer diagnostic capabilities for the secondary air-injection system. Consult the service manual for specific models. Catalytic converter efficiency is monitored on OBD-II compliant vehicles and will set DTCs if the converter becomes degraded. Follow manufacturer-specific test routines if a catalyst DTC is set.

3. Inspect, test, service, and replace mechanical components and electrical/electronically-operated components and circuits of secondary air injection systems.

Check all hoses and pipes in the system for looseness and rusted or burned conditions. Burned or melted air-injection hoses or valve indicate leaking check valves. Inspect the check valves and replace them if they show signs of leakage. With the engine idling, listen for noises from the pump (if equipped). Check the air pump drive belt; adjust it if loose, or replace it if worn or damaged. Check for adequate airflow from the pump and test for airflow to the exhaust manifold during engine warm-up and for flow to the catalytic converter when the fuel system enters closed loop.

A properly operating air pump should raise tailpipe oxygen readings above 2 percent and often show levels as high as 3 to 8 percent. Airflow should also divert to the air cleaner or atmosphere when the engine is decelerated rapidly or during high RPM operation. This is the function of the air pump diverter valve. The air-injection switching valve controls airflow to the exhaust manifolds or catalytic converter depending upon engine operating conditions. On some computer-controlled vehicles, the PCM can control the diverter and switching valves according to engine operating conditions.

The computer commands to the air-injection system valves can be checked on a scan tool. A scan tool may allow testing on some electric air pumps by means of allowing the technician to turn the air pump on and off. Check the vehicle service manual to identify system components and determine proper test procedures.

4. Inspect and test the catalytic converter(s); interpret catalytic converter related DTCs; analyze related scan tool data to determine root cause of DTCs; determine needed repairs.

If the catalytic converter rattles when tapped with a soft hammer, then the internal components are loose and the converter should be replaced. When a catalytic converter is restricted, a significant loss of power and limited top speed will be noticed.

Various tests are available to determine if the converter is functional. One test is to measure converter inlet and outlet temperatures with a temperature probe to see if the converter lights off. A properly operating converter should show a temperature increase at the outlet compared to the inlet. Most manufacturers call for a 10 percent increase in temperature. A temperature increase greater than several hundred degrees could indicate the converter is working too hard due to excessive hydrocarbons present in the exhaust. The engine should be checked for misfiring. A technician can use an exhaust analyzer to perform a cranking CO_2 test, an oxygen storage test, or take measurements before and after the converter to calculate converter efficiency, also called an intrusive test.

Intrusive converter testing should produce efficiency results above 80 percent. The oxygen storage test should show less than a 1.2 to 1.7 percent increase in oxygen readings during snap-throttle acceleration from 2,000 rpm if the converter is working properly and storing oxygen. A cranking CO_2 test determines if a pre-heated converter can convert hydrocarbons to carbon dioxide while the engine is cranked with the ignition system disabled. The fuel system must be supplying fuel because that is the source of the hydrocarbons. Tailpipe CO_2 reading should be above 12 percent and hydrocarbons should stay below 500 ppm.

4. Evaporative Emissions Controls (3 Questions)

1. Test and diagnose emissions or driveability problems caused by the evaporative emissions control system.

The evaporative emissions control (EVAP) system captures and stores vapors from the fuel system in a charcoal canister to be burned once the engine is started. These vapors are purged from the canister by engine vacuum after the engine is started and run off idle. If the engine purges vapors from the charcoal canister during idle, then a rich condition and possible rough idle could result. Fuel-saturated charcoal canisters may cause excessively rich air/fuel ratios during acceleration, which may cause state emission test failures. Leaks in the evaporative emission system can cause customer complaints of gasoline odors in or around the vehicle.

Some later-model OBD II vehicles have an enhanced EVAP system that can monitor purge flow rate as well as determine if the system has fuel vapor leaks. Problems with system purging or leaks will set diagnostic trouble codes on these systems.

2. Interpret evaporative emissions-related scan tool data and DTCs; determine needed repairs.

Once diagnostic trouble codes are retrieved, the technician needs to consult the service manual for proper diagnostic procedures and system operation. Trouble codes will identify whether the problem is with canister fuel vapor purging or system leaks. Due to the fact that the PCM can identify very small leaks and store a trouble code, special test equipment may be needed to diagnose and locate the problem.

3. Inspect, test, and replace canister, lines/hoses, mechanical and electrical components of the EVAP control systems.

A careful visual inspection of the evaporative system should be performed when a customer complains of gasoline odors. A gas analyzer or smoke machine will help identify any leaks. Often hoses will be damaged or left disconnected after other repairs are done. EVAP system purge and vent solenoids need to be checked for proper resistance and electrical operation as well as proper mechanical operation.

Most purge solenoids are normally closed and block engine vacuum to the canister when off. Energizing the purge solenoid will allow vacuum through to purge the canister. Vent solenoids are normally open and allow air to pass through the solenoid until the solenoid is energized. The vent solenoid is used to seal the system so the PCM can test for system leaks. Charcoal canisters may be equipped with filters that should be replaced at the manufacturer's specified interval.

The fuel tank cap should be carefully inspected for proper application and sealing if a system leak code is set. This is one of the most common problems setting EVAP system leak codes. Special testers are available to test the pressure and vacuum valves in the gas cap. Fuel tanks may also be equipped with a rollover valve to prevent fuel from escaping through the evaporative system if an accident caused the vehicle to turn upside down or rollover.

Some evaporative systems may use a tank pressure control valve (TPCV) that controls the flow of vapors to the charcoal canister. If fuel vapor pressure in the fuel tank is below 1.5 in. Hg, then the valve will be closed and fuel vapors will be stored in the tank. When vapor pressure exceeds the set point of the valve, the vapors are vented to the charcoal canister. The TPCV also provides vacuum relief to protect against vacuum buildup in the fuel tank.

E. Computerized Engine Controls Diagnosis and Repair (including OBD II) (13 Questions)

1. Retrieve and record DTCs for OBD II monitor status and freeze frame data, if applicable.

Retrieving diagnostic trouble codes varies greatly among the many different manufacturers' vehicles. Manual code retrieval on pre-OBD II vehicles include some of the following procedures: installing a jumper wire across the proper terminals of a diagnostic connector and counting lamp flashes, cycling the ignition switch on and off three times in a five-second period to signal the PCM to enter diagnostic mode and counting lamp flashes, or turning a switch on the PCM to enter diagnostics and counting the blinking of LED's in the PCM. Many pre-OBD II vehicles also support scan tool code retrieval. OBD II vehicles require the use of a scan tool to retrieve and clear diagnostic trouble codes. The scan tool will also allow viewing freeze frame data that is stored in the PCM when a diagnostic trouble code is set.

2. Diagnose the causes of emissions or driveability problems resulting from failure of computerized engine controls with stored or active DTCs.

Once a fault has been detected by the PCM, it stores a DTC in memory, and if the fault affects exhaust emissions, then it will light the malfunction indicator lamp (MIL). The technician then retrieves the stored DTC and accesses a diagnostic flow chart. The diagnostic flow chart leads the technician through a series of steps to determine the actual problem.

When diagnosing a fault, it is useful for the technician to be aware of the circumstances that cause the control module to set a fault. There are a specific set of circumstances that cause the control module to set a fault. The information found in the workshop manual will allow the technician to further refine the diagnostics necessary to solve the problem.

An important step in the process of diagnosing computer trouble codes is to determine if the code is a history (memory) code or if the code is current, which means the fault is present at the time the trouble code was retrieved. Following a trouble code flow chart, a history code can cause misdiagnosis and replacement of unnecessary parts. Most late-model PCMs will report codes as either current or history. On early model systems that do not differentiate between current and history codes, the code should be cleared and the vehicle driven to see if the code resets. On OBD II systems, the codes should not be cleared until the vehicle is repaired to prevent the PCM from erasing stored freeze frame data.

3. Diagnose the causes of emissions or driveability problems resulting from failure of computerized engine controls with no DTCs.

Vehicles with computerized engine control systems may exhibit many driveability or emissions problems without setting a diagnostic trouble code. A hesitation on acceleration can be experienced from a faulty throttle position sensor (TPS). A bad spot in the sensor circuit board could be causing the signal voltage to momentarily drop. The computer interprets this as a decrease in throttle position, when actually the vehicle is still accelerating. The computer may never set a DTC based on this type of fault, because the voltage never varies above or below the voltage levels needed to set a TPS code.

A leak in the vacuum hose to a manifold absolute pressure sensor (MAP) may cause a higher than normal reading from the MAP sensor. The computer interprets the reading as higher engine load and increases fuel delivery. Higher exhaust emissions will result, but a DTC may not be stored, because the MAP sensor voltage output is still within the normal operational range.

These conditions require careful pinpoint testing by the technician to identify the root cause of the problem. An understanding of normal system readings and specifications is needed so the technician can diagnose problems that do not set trouble codes in the computer. Specialty tools such as power graphing multimeters, lab scopes with record modes, or graphing scan tools make the job of finding intermittent circuit or sensor problems much easier.

4. Use a scan tool, DMM, or DSO to inspect or test computerized engine control system sensors, actuators, circuits, and powertrain/engine control module (PCM/ECM); determine needed repairs.

In order to control engine operation the PCM must have a certain set of sensor inputs on which to make decisions. Once these inputs are received, the PCM processes the signals and decides which course of action to take. The PCM then outputs signals to a series of actuators that in turn provide the engine with the things it needs to operate efficiently. In order to troubleshoot a system effectively, the technician must confirm that the sensor is transmitting the appropriate signal to the PCM and the PCM is transmitting the proper signal to the actuator.

To properly diagnose today's vehicles, a technician must be familiar with and be able to operate sophisticated diagnostic equipment such as various scan tools and DSOs. The scan tool allows a convenient means of accessing computer sensor data as well as the output commands or status of the engine control system. Many scan tools today allow bi-directional testing where the technician can take direct control of items like idle speed, or perform testing such as engine power balance, by disabling individual fuel injectors. DMMs and DSOs allow detailed circuit and component testing.

Computer and sensor power feeds; grounds and signal wires can be tested with a DMM. Sensor and actuator waveform analysis is performed with a DSO. Items such as fuel injectors and oxygen sensors are tested most effectively through waveform analysis. Consult equipment and manufacturer's test procedures to utilize these test instruments to their fullest capabilities.

5. Measure and interpret voltage, voltage drop, amperage, and resistance using DMM readings.

Using a DMM, the technician can evaluate circuit integrity by testing available voltage, voltage drop, amperage, and resistance readings. Available voltage tests confirm if a component is receiving the proper amount of voltage. A voltage drop test is used to locate unwanted circuit resistance when the circuit is energized.

Amperage tests can locate shorted or high resistance actuators. Resistance measurements can be used to compare components to specifications or for testing wiring harness continuity. Connecting a DMM to measure available voltage is done by making a parallel connection across the circuit. The read lead is connected at the desired test point and the black lead is connected to the negative battery terminal or engine block. When a voltage drop test is performed, the DMM leads are connected across a component or section of a circuit on the same side of the circuit, either the power or ground side of the circuit.

The reading on the voltmeter indicates the amount of voltage used by the component or section of the circuit between the meter leads. Current must be flowing in the circuit for a voltage drop to occur. A high voltage drop across a conductor or connector indicates excessive resistance. A low voltage drop across a load means resistance is present elsewhere in the circuit. Voltage drop testing across the power and ground sides of a circuit will identify the location of the unwanted resistance.

Inserting the meter in series with the circuit performs amperage testing with a DMM. All DMMs are rated for the amount of current they can measure directly. Care must be taken to prevent connecting the meter into a circuit with higher current draw than the meter can measure. Meters are protected with an internal fuse that will open and require replacement if the current rating of the meter is exceeded. Current probes are available that will allow a DMM to make high current measurements. Resistance measurements can be made on components that are removed from the circuit. Whenever resistance tests are performed on circuit wiring, battery power must be removed from the circuit under test. All ohmmeters are self-powered and must not be used on live circuits.

6. Test, remove, inspect, clean, service, and repair or replace power and ground distribution circuits and connections.

The *power distribution circuit* is the power and ground circuits from the battery, through the ignition switch and fuses, to the individual circuits on the vehicle. Connections must be free of corrosion, as it adds unwanted resistance to current flow.

7. Practice recommended precautions when handling static-sensitive devices and/or replacing the PCM. Remove and replace the PCM/ECM; reprogram as needed.

The PCM must be handled with care when replacing to ensure no static discharge into the computer. Locate a good ground on the vehicle and connect a grounding strap from yourself to the vehicle ground before installing the component. Most modern-day computers require flash programming to upload new operating parameters for smooth operation. A battery charger should be installed to ensure no battery loss when reprogramming a PCM.

There are many static-sensitive components used on vehicles including powertrain, body, transmission, and ABS control computers as well as electronic instrument clusters. Static sensitive components are shipped in an antistatic envelope. This envelope should not be opened until you are ready to install the component. Locate a good ground on the vehicle and connect a grounding strap from you to the vehicle ground before installing the component. Do not handle the component unnecessarily, and do not move around on the vehicle seat when installing the component.

8. Diagnose driveability and emissions problems resulting from failures of interrelated systems (such as cruise control, security alarms/theft deterrent, torque controls, traction controls, torque management, A/C, non-OEM installed accessories).

Today's vehicles have multiple computers with multiple functions. The computers have the ability to communicate with each other. One computer receives some sensor inputs, and the signal is forwarded to other computers. If this signal is not received, then this can be interpreted as an incorrect input and may cause output problems from the processor. Theft deterrent systems may cause stalling or no-start conditions that can be difficult to trace or lead a technician toward performing unnecessary testing and parts replacement. This may be especially true with aftermarket alarm systems. Often when control modules that communicate with each other like the PCM and theft deterrent module are replaced, re-learn procedures may need to be performed before normal operation will occur.

Other driveability concerns can occur from problems in the cruise control or traction control systems, such as surging or loss of power. Electrical interference may result from certain aftermarket-installed accessories, such as stereo amplifiers. Improper installation or wiring damage is also a concern when non-OEM components are installed.

A technician must identify all systems that are used on a vehicle and what possible interaction they may have on the PCM. The vehicle service manual and service bulletins should be consulted when problems are suspected in interrelated systems.

9. Diagnose the causes of emissions or driveability problems resulting from computerized spark timing controls; determine needed repairs.

The ignition module receives an input from a Hall-effect pickup or a variable reluctance sensor; this signal is used to fire the coil(s) on start up. The ignition module sends this signal to the PCM, and the PCM interprets it as an RPM input. This signal between the ignition module and PCM is a digital signal. The PCM then sends a varying digital signal back to the ignition module. The module uses this signal as a computed timing signal and fires the coil(s) based on this information.

Problems with spark timing controls will often set diagnostic trouble codes. If the computer detects a spark timing control circuit problem, the engine will operate at base ignition timing, and a lack of power may result. Improper sensor inputs may cause changes in spark timing control and lead to increased emissions. For example, a coolant sensor that is out of calibration and sends a colder-than-actual temperature reading may cause the PCM to increase spark timing. This in turn will increase fuel delivery and spark advance, resulting in excessive fuel consumption and emissions.

10. Verify the repair and clear the DTCs; run all OBD II monitors, and verify the repair.

Prior to the introduction of OBD II, each manufacturer had its own method for erasing DTCs from the memory of a PCM. These procedures should always be followed. Normally, verification of the repair is done by operating the engine and the related system and checking to see if the operation triggered the DTC. If it did not, the problem was probably solved.

On OBD II equipped vehicles, the fail records and the freeze frame data for the DTC that was diagnosed should be reviewed and recorded. Then use a scan tool's clear DTCs or clear info functions to erase the DTCs from the PCM's memory. Operate the vehicle within the conditions noted in the fail records and/or the freeze frame data. Then monitor the status information for the specific DTC until the diagnostic test associated with that DTC runs.

Sample Preparation Exams

INTRODUCTION

Included in this section are a series of six individual preparation exams that you can use to help determine your overall readiness to successfully pass the Engine Performance (A8) ASE certification exam. In Section 7 of this book, you will find blank answer sheet forms you can use to designate your answers to each of the preparation exams. Using these blank forms will allow you to attempt each of the six individual exams multiple times without risk of viewing your prior responses.

Upon completion of each preparation exam, you can determine your exam score using the answer keys and explanations located in Section 6 of this book. Included in the explanation for each question is the specific task area being assessed by that individual question. This additional reference information may prove useful if you need to refer back to the task list located in Section 4 for additional support.

PREPARATION EXAM 1

1. Technician A says valve adjustment should always be performed on a cold engine. Technician B says the piston should be placed at top dead center (TDC) of the compression stroke. Who is correct?

   A. A only
   B. B only
   C. Both A and B
   D. Neither A nor B

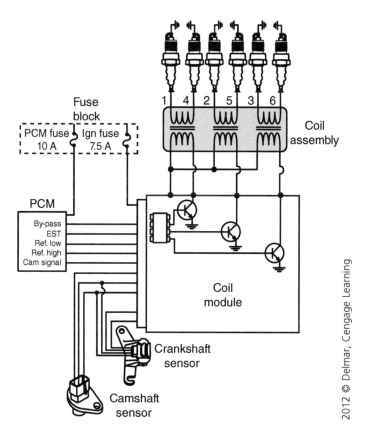

2012 © Delmar, Cengage Learning

2. Refer to the illustration. Technician A says a shorted coil could affect two cylinders. Technician B says the schematic shows a waste spark ignition system. Who is correct?

    A. A only

    B. B only

    C. Both A and B

    D. Neither A nor B

3. Technician A says fuel pressure readings that are above specifications could be caused by a stuck-open fuel pressure regulator. Technician B says if fuel is present at the vacuum hose port of a fuel pressure regulator, then the regulator is stuck open. Who is correct?

    A. A only

    B. B only

    C. Both A and B

    D. Neither A nor B

4. Technician A says if the positive crankcase ventilation (PCV) valve rattled when shaken, then the PCV system is OK. Technician B says the PCV system vents excess pressure formed in the crank case from piston blow-by. Who is correct?

    A. A only

    B. B only

    C. Both A and B

    D. Neither A nor B

5. Technician A says a greenish corrosion on terminals results in high resistance in the circuit. Technician B says loose retaining lock tabs on a terminal can cause high resistance in connector terminals. Who is correct?

   A. A only

   B. B only

   C. Both A and B

   D. Neither A nor B

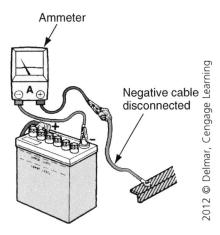

Ammeter

Negative cable disconnected

2012 © Delmar, Cengage Learning

6. Technician A says a key-off current draw test is being performed in the above illustration. Technician B says the allowable reading for this test is less than .05 amps. Who is correct?

   A. A only

   B. B only

   C. Both A and B

   D. Neither A nor B

7. Technician A says a single-cylinder misfire diagnostic trouble code can be caused by a defective coil on some vehicles. Technician B says a low fuel-pump pressure can cause a single-cylinder misfire diagnostic trouble code. Who is correct?

   A. A only

   B. B only

   C. Both A and B

   D. Neither A nor B

8. Technician A says a vehicle with a restricted fuel filter can still have fuel pressure within the specifications. Technician B says if a restricted fuel filter is suspected, then a fuel pump volume test can be performed. Who is correct?

   A. A only

   B. B only

   C. Both A and B

   D. Neither A nor B

9. All of the following can cause the engine to spark knock EXCEPT:

    A. A stuck-open exhaust gas recirculation (EGR) valve

    B. A stuck-closed EGR

    C. A broken vacuum hose going to the EGR

    D. A defective cooling system thermostat

10. Technician A says when replacing the powertrain control module (PCM), the new PCM may have to be reprogrammed in order for it to operate. Technician B says some drivability problems can be fixed by reprogramming the PCM. Who is correct?

    A. A only

    B. B only

    C. Both A and B

    D. Neither A nor B

11. Technician A says one of the first steps in diagnosing a drivability complaint is to verify the driver's complaint. Technician B says one of the first steps in diagnosing a drivability complaint is to perform a thorough visual inspection. Who is correct?

    A. A only

    B. B only

    C. Both A and B

    D. Neither A nor B

12. Technician A says the normal secondary-circuit resistance of an ignition coil is low (5 ohms or less). Technician B says the spark plug wire is a secondary component that should have a resistance of 20,000 ohms or more per foot. Who is correct?

    A. A only

    B. B only

    C. Both A and B

    D. Neither A nor B

13. Technician A says a vacuum leak can occur under the intake manifold and cause oil consumption. Technician B says propane is a good method for locating vacuum leaks. Who is correct?

    A. A only

    B. B only

    C. Both A and B

    D. Neither A nor B

14. Technician A says a secondary air-injection system directs output from the air pump to the exhaust manifold during engine warm-up and switches air to the catalytic converter during closed-loop operation. Technician B says air injection will have little or no effect on tailpipe carbon dioxide readings. Who is correct?

    A. A only

    B. B only

    C. Both A and B

    D. Neither A nor B

15. Technician A says the best way to clear diagnostic trouble codes is to remove the battery negative terminal for five seconds. Technician B says the monitor for the repaired system should be performed before the vehicle is returned to the customer. Who is correct?

    A. A only
    B. B only
    C. Both A and B
    D. Neither A nor B

16. All of the following are true of the cylinder leakage test EXCEPT:

    A. Air loss and bubbles in the radiator indicate a bad intake valve guide.
    B. Air loss from the oil filler cap indicates worn piston rings.
    C. A gauge reading of 0 percent indicates no cylinder leakage.
    D. Air loss from the exhaust indicates a valve problem.

17. Technician A says a faulty cam shaft position sensor can cause a no-start condition. Technician B says some vehicles will start without an operating camshaft position sensor (CMP). Who is correct?

    A. A only
    B. B only
    C. Both A and B
    D. Neither A nor B

18. Technician A says the exhaust can be checked for restrictions using a vacuum gauge. Technician B says the back pressure in the exhaust should not exceed 1.5 pounds per square inch (PSI) at idle. Who is correct?

    A. A only
    B. B only
    C. Both A and B
    D. Neither A nor B

19. Technician A says a leaking gas cap gasket can cause an evaporative emissions failure. Technician B says some evaporative emissions canisters have a replaceable filter. Who is correct?

    A. A only
    B. B only
    C. Both A and B
    D. Neither A nor B

20. Technician A says a diagnostic trouble code (DTC) tells you which component is malfunctioning. Technician B says if a fault exists that affects emissions, then the malfunction indicator lamp (MIL) will be illuminated. Who is correct?

    A. A only
    B. B only
    C. Both A and B
    D. Neither A nor B

21. Technician A says blue smoke in the exhaust indicates oil being burned in the combustion chamber. Technician B says white smoke in the exhaust indicates a rich air/fuel ratio. Who is correct?

    A. A only

    B. B only

    C. Both A and B

    D. Neither A nor B

22. All of the following are components of the primary circuit EXCEPT:

    A. Ignition coil

    B. Ignition switch

    C. Spark plug wire

    D. Battery

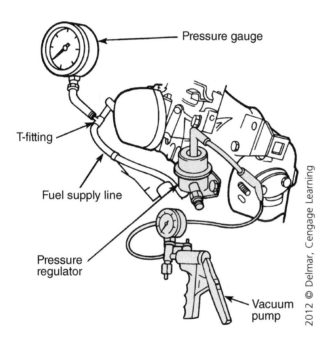

23. Refer to the above illustration. Technician A says fuel pressure regulator operation is being tested. Technician B says as vacuum is applied to the regulator, the fuel pressure should increase. Who is correct?

    A. A only

    B. B only

    C. Both A and B

    D. Neither A nor B

24. Technician A says the PCV valve flow is high at idle. Technician B says PCV valve flow is high at cruising speed on a vehicle with an 8-cylinder engine. Who is correct?

    A. A only

    B. B only

    C. Both A and B

    D. Neither A nor B

25. A vehicle with a MIL illuminated is being diagnosed. The DTC stored in the PCM is PO300 (Random/Multiple-Cylinder Misfire Detected). Technician A says this is a one-trip failure, because catalyst damage can occur. Technician B says the MIL will flash on this type of code. Who is correct?

     A. A only
     B. B only
     C. Both A and B
     D. Neither A nor B

26. Technician A says that cranking the engine with the throttle fully depressed will force a lean mixture to clear a flooded engine on some vehicles. Technician B says that high system voltage will increase fuel injector on time (pulse width). Who is correct?

     A. A only
     B. B only
     C. Both A and B
     D. Neither A nor B

27. Which of the following is the most likely cause of ignition coil failure?

     A. Prolonged open circuit in the secondary
     B. Open ignition module primary circuit
     C. Short-to-ground in the secondary
     D. Open in the trigger device circuit

28. Technician A says performing a fuel pressure test confirms proper operation of the fuel injector. Technician B says it is possible to have an electrical problem with an injector, even though the fuel pressure drop is within specifications. Who is correct?

     A. A only
     B. B only
     C. Both A and B
     D. Neither A nor B

29. Technician A says the EGR system is used to raise combustion chamber temperature. Technician B says that EGR systems that use an EGR valve position sensor should read about .7 volts with the EGR valve at full open. Who is correct?

     A. A only
     B. B only
     C. Both A and B
     D. Neither A nor B

30. A scan test of the computer system on a late-model fuel-injected engine reveals a bank #1 long-term fuel trim value of -19, and a bank #2 long-term fuel trim value of -18 with the engine idling. Technician A says these readings could be caused by a defective fuel pressure regulator. Technician B says a restricted fuel return line could cause these readings. Who is correct?

     A. A only
     B. B only
     C. Both A and B
     D. Neither A nor B

31. Technician A says a stethoscope can be used to pinpoint engine noises. Technician B says you may use a long screwdriver for noise diagnosis if a stethoscope is not available. Who is correct?

    A. A only
    B. B only
    C. Both A and B
    D. Neither A nor B

32. Engine oil does all of the following tasks EXCEPT:

    A. Controls the variable valve timing system
    B. Helps cool the charging system
    C. Helps seal the piston rings
    D. Lubricates the cam phaser

33. Technician A says that turning off fuel injectors at high RPM is the purpose of the rev limiter—to protect the engine from damage or limit vehicle speed. Technician B says that turning off fuel injectors while the engine is running must be done at speeds above 45 mph. Who is correct?

    A. A only
    B. B only
    C. Both A and B
    D. Neither A nor B

34. Technician A says to use an anti-backfire valve to prevent backfiring during deceleration on pump-driven air-injection systems. Technician B says to prevent exhaust gases from back-flowing into the air-injection control valves or pump, check valves can be used in the exhaust manifold and converter feed pipes. Who is correct?

    A. A only
    B. B only
    C. Both A and B
    D. Neither A nor B

35. A mass air flow (MAF) load-calculating-type port fuel-injected engine runs fine at idle, but hesitates under acceleration. No DTCs are stored. Technician A says to check for a restricted MAF inlet screen. Technician B says a bad spot in the throttle position sensor signal could be the cause. Who is correct?

    A. A only
    B. B only
    C. Both A and B
    D. Neither A nor B

36. Which of the following is not a cause of a noisy valve train?

    A. High oil pressure
    B. Collapsed lifter
    C. Incorrect valve adjustment
    D. Bent pushrod

37. A cylinder misfire will cause all the following EXCEPT:

    A. High hydrocarbon emissions
    B. High oxygen level in the exhaust
    C. Damage to the catalytic converter
    D. High carbon dioxide readings

38. A vehicle is being diagnosed for a PO134 DTC (Oxygen Sensor Circuit, No Activity Detected) (bank #1, sensor #1). Technician A says the problem could be high fuel system pressure. Technician B says the problem is more likely confined to the downstream oxygen sensor on the bank containing cylinder #1. Who is correct?

    A. A only
    B. B only
    C. Both A and B
    D. Neither A nor B

39. Technician A says that an EGR valve stuck closed will cause detonation. Technician B says that an EGR valve stuck closed will cause high NOx emissions. Who is correct?

    A. A only
    B. B only
    C. Both A and B
    D. Neither A nor B

40. Technician A says that the resistance in the power side of the power distribution circuit cannot be over 200 ohms of resistance. Technician B says the power side of the power distribution circuit can be voltage-drop tested to check the circuit for resistance. Who is correct?

    A. A only
    B. B only
    C. Both A and B
    D. Neither A nor B

41. If the vacuum drops slowly to a low reading when a vacuum gauge is connected to the intake manifold and the engine is accelerated and held at a steady speed, then which of the following is the most likely cause?

    A. A rich fuel mixture
    B. Over-advanced ignition timing
    C. Sticking valves
    D. A restricted exhaust

42. A vehicle with DTC code PO304 (Cylinder #4 Misfire) is being diagnosed. Technician A says high fuel pump pressure could be the cause. Technician B says an open spark plug wire could be the cause. Who is correct?

    A. A only
    B. B only
    C. Both A and B
    D. Neither A nor B

43. Which of the following is the LEAST LIKELY cause of a fuel tank leak?

   A. Rust

   B. Road damage

   C. Retaining straps

   D. Leaking tank seams

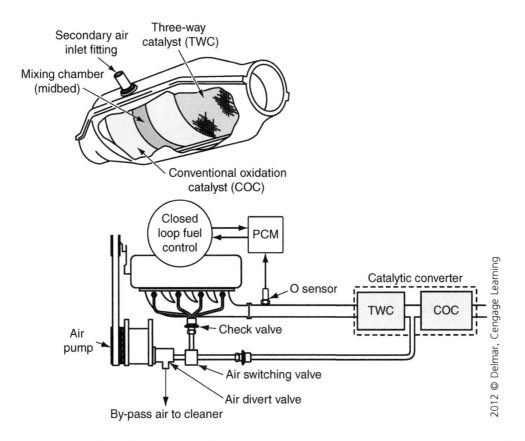

44. Refer to the above illustration. Technician A says the air pump should pump the air into the exhaust manifold when in open loop. Technician B says the air should be pumped into the catalytic converter once the vehicle enters closed loop. Who is correct?

   A. A only

   B. B only

   C. Both A and B

   D. Neither A nor B

45. A vehicle has a DTC for an open vehicle speed sensor circuit. Which of the following systems would most likely be affected?

   A. Air conditioning system

   B. Manual transmission shift points

   C. Traction control system (TCS)

   D. Cruise control system

46. Technician A says the electrolyte level is important in a non-serviceable battery. Technician B says on some batteries, the electrolyte level can be checked in a sealed battery by looking through the translucent battery case. Who is correct?

    A. A only
    B. B only
    C. Both A and B
    D. Neither A nor B

47. What term describes the condition when the air/fuel mixture is ignited before the spark plug is fired?

    A. Dieseling
    B. Pre-ignition
    C. Lean ramping
    D. Early timing

48. Technician A says that special fuel injection hose must be used when replacing the fuel lines on a fuel-injected vehicle. Technician B says the fuel supply hose is usually smaller than the fuel return hose. Who is correct?

    A. A only
    B. B only
    C. Both A and B
    D. Neither A nor B

49. Technician A says oil found inside the air filter housing may be caused by excessive crankcase blow-by. Technician B says a plugged PCV valve will cause excessive crankcase pressure. Who is correct?

    A. A only
    B. B only
    C. Both A and B
    D. Neither A nor B

50. A DTC for an intake air temperature sensor may be set by any of the following conditions EXCEPT:

    A. An open in the voltage reference wire
    B. Operating the vehicle in extremely cold climates
    C. A short in the voltage reference wire
    D. An out-of-range sensor input

PREPARATION EXAM 2

1. Which of the followings is the most likely cause of blue exhaust smoke?

   A. Worn piston rings

   B. Worn piston pins

   C. Rich fuel mixtures

   D. Internal coolant leak

2. All of the following components are part of the fuel pump control circuit EXCEPT:

   A. Fuel pump relay

   B. Powertrain control module (PCM)

   C. Ignition switch

   D. Body control module (BCM)

3. An EGR vacuum regulator solenoid (EGRV) is thought to be inoperative. Technician A says when measuring the resistance of the solenoid, a lower-than-specified reading means the windings are open. Technician B says an infinite reading means the winding is shorted. Who is correct?

   A. A only

   B. B only

   C. Both A and B

   D. Neither A nor B

4. Technician A says most computer inputs are received from other computers. Technician B says one input might affect other computers. Who is correct?

   A. A only

   B. B only

   C. Both A and B

   D. Neither A nor B

5. A cylinder-power balance test on a distributor-ignition vehicle can indicate all of the following problems EXCEPT:

   A. Burnt valve

   B. Cracked piston

   C. Blown head gasket

   D. Weak ignition coil

6. While testing the secondary ignition with an oscilloscope, which of the following is the LEAST LIKELY cause of high resistance in the ignition secondary circuit?

   A. A burnt rotor button

   B. Corroded ignition coil terminal

   C. Wide spark plug gap

   D. No dielectric grease on the ignition module mounting surface

7. Technician A says some manufacturers do not allow the cleaning of a throttle body with throttle body cleaner. Technician B says a buildup of gum and carbon deposits may cause rough idle operation. Who is correct?

    A. A only
    B. B only
    C. Both A and B
    D. Neither A nor B

8. Technician A says with the PCV valve disconnected from the rocker cover, there should be vacuum at the valve with the engine idling. Technician B says when the PCV valve is removed and shaken, there should not be a rattling noise. Who is correct?

    A. A only
    B. B only
    C. Both A and B
    D. Neither A nor B

9. Technician A says that if an emissions-related diagnostic trouble code (DTC) is set, then a freeze frame is also stored. Technician B says if an emissions-related DTC is set, then the malfunction indicator lamp will illuminate. Who is correct?

    A. A only
    B. B only
    C. Both A and B
    D. Neither A nor B

10. Which of the following is most likely to cause a double-knocking noise with the engine at an idle?

    A. Excessive main bearing thrust clearance
    B. Worn main bearing
    C. Worn piston wrist pins
    D. Excessive rod bearing clearance

11. Technician A says the plug wire resistance should not exceed 10 megohms of resistance. Technician B says plug wire insulation can be checked with a saltwater solution in a spray bottle. Who is correct?

    A. A only
    B. B only
    C. Both A and B
    D. Neither A nor B

12. Technician A says a pressure drop test can be performed to test for a restricted fuel filter. Technician B says some vehicles do not use an in-line filter; the fuel filter is in the fuel tank. Who is correct?

    A. A only
    B. B only
    C. Both A and B
    D. Neither A nor B

13. Technician A says that *enabling criteria* are specific conditions that must be met before a monitor will run, such as coolant temperature or engine speed. Technician B says that *pending conditions* are conditions that exist that prevent a specific monitor from running, such as an oxygen sensor fault code preventing an oxygen sensor heater monitor from running. Who is correct?

    A. A only
    B. B only
    C. Both A and B
    D. Neither A nor B

14. Technician A says that the first step of any diagnostic procedure is to check for DTCs. Technician B says the search for technical service bulletins (TSB) should be consulted before any repairs are made. Who is correct?

    A. A only
    B. B only
    C. Both A and B
    D. Neither A nor B

15. A primary ignition circuit on a vehicle checks good, but there is no spark from the spark plug wire. There is spark at the coil wire. This could be caused by any of the following:

    A. Bad ignition coil
    B. Bad ignition module
    C. Shorted rotor button
    D. Broken distributor gear

16. Technician A says the wastegate on a turbo charger closes to control boost by redirecting exhaust gasses around the turbine. Technician B says once boost pressure is under control, the wastegate opens completely. Who is correct?

    A. A only
    B. B only
    C. Both A and B
    D. Neither A nor B

17. All of the following are part of the secondary air system EXCEPT:

    A. Exhaust check valve
    B. Diverter valve
    C. Back pressure transducer
    D. Air pump

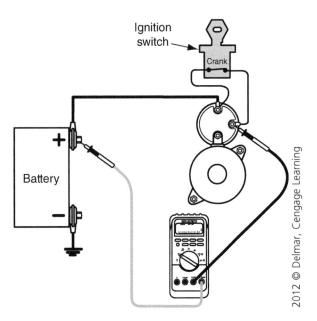

18. Technician A says an ignition switch starter-control circuit voltage drop test is being performed in the above illustration. Technician B says a reading of 2 volts is acceptable on this circuit. Who is correct?

    A. A only
    B. B only
    C. Both A and B
    D. Neither A nor B

19. Technician A says that during a cylinder compression test, low readings on adjacent cylinders may be caused by a cracked cylinder head. Technician B says a low reading on a single cylinder that increases when a tablespoon of oil is added is probably a valve problem. Who is correct?

    A. A only
    B. B only
    C. Both A and B
    D. Neither A nor B

20. Technician A says when testing the ignition coil, both the primary and secondary winding should be checked for resistance. Technician B says maximum coil output testing can be performed with an oscilloscope. Who is correct?

    A. A only
    B. B only
    C. Both A and B
    D. Neither A nor B

21. A seized-up turbo charger can cause vacuum readings to do which of the following?

    A. Show a continuous gradual drop as engine speed is increased
    B. Fluctuate between 15 and 21 inches at an idle
    C. Drop off about 3 inches at an idle
    D. Drop off about 6 inches at an idle

22. Which of the follow diagnostic trouble codes are the highest priority codes and should be diagnosed first?

    A. Transmission trouble codes
    B. Emissions-related trouble codes
    C. Fuel-related trouble codes
    D. Misfire-related trouble codes

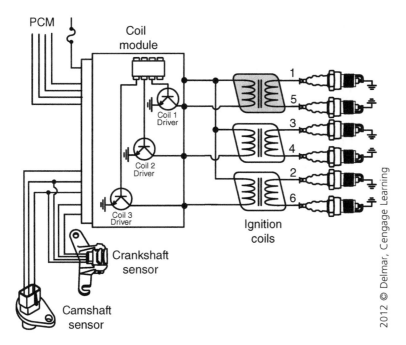

23. A vehicle with the ignition system illustrated has a no spark condition on cylinders #1 and #5. Technician A says the ignition module could have a bad coil driver. Technician B says the crankshaft sensor could have an open circuit, causing the no spark at #1 and #5. Who is correct?

    A. A only
    B. B only
    C. Both A and B
    D. Neither A nor B

24. All of the following could cause high starter current readings EXCEPT:

    A. A short circuit
    B. An open circuit
    C. Tight engine crankshaft bearings
    D. A defective starter

25. A vehicle with sequential fuel injection (SFI) has high fuel-pump pressure at idle. Which of the following could be the cause?

    A. Excessive manifold vacuum
    B. A leaking fuel pump drain-back check valve
    C. Low manifold vacuum
    D. A stuck-closed fuel injector

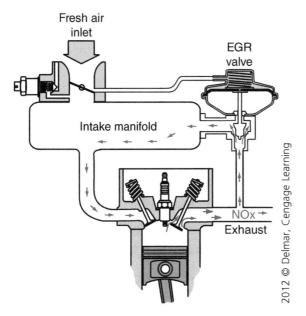

Fresh air inlet

EGR valve

Intake manifold

NOx →

Exhaust

2012 © Delmar, Cengage Learning

26. Technician A says the exhaust gas recirculation valve (EGR) illustrated allows engine exhaust to enter the intake manifold of the engine. Technician B says the EGR system lowers combustion chamber temperature. Who is correct?

    A. A only
    B. B only
    C. Both A and B
    D. Neither A nor B

27. Technician A says a scan tool or code reader is used to retrieve DTCs from an OBD II vehicle. Technician B says the DTCs can be retrieved by watching and counting the number of flashes from the malfunctioning indicator lamp (MIL) on OBD II vehicle. Who is correct?

    A. A only
    B. B only
    C. Both A and B
    D. Neither A nor B

28. A multi-trace oscilloscope can test all of the following EXCEPT:

    A. Throttle position sensor
    B. Mass air flow (MAF) sensor
    C. Crankshaft position (CKP) sensor
    D. Air/fuel ratio

29. Technician A says the secondary ignition circuit is designed to handle high voltages up in the thousand volts. Technician B says the secondary circuit produces dangerously high amperage. Who is correct?

    A. A only
    B. B only
    C. Both A and B
    D. Neither A nor B

30. A vehicle has a low power complaint with a hissing sound coming from under the vehicle at wide-open throttle (WOT). Technician A says the exhaust back pressure should be checked. Technician B says the manifold vacuum should be checked. Who is correct?

    A. A only
    B. B only
    C. Both A and B
    D. Neither A nor B

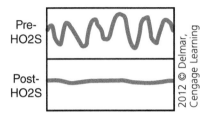

31. A multi-trace oscilloscope is being used on the pre- and post-catalytic converter heated oxygen sensors, and the results are shown as illustrated. Technician A says the catalytic converter is working correctly. Technician B says the post-converter heated O_2 sensor should be replaced. Who is correct?

    A. A only
    B. B Only
    C. Both A and B
    D. Neither A nor B

32. Which of the following conditions is most likely caused by a manifold absolute pressure (MAP) sensor?

    A. Poor fuel economy
    B. Excessive idle speeds
    C. Spark knock
    D. Erratic speedometer operation

33. What behavior should a viscous-drive fan clutch exhibit when rotated by hand with the engine off?

    A. No resistance
    B. More resistance cold
    C. More resistance hot
    D. Not freewheel any

34. When installing and timing the distributor, Technician A says the engine must be timed referencing TDC on the specified cylinder's exhaust stroke. Technician says if the engine is timed on the compression stroke, then the distributor will be 180 degrees off. Who is correct?

    A. A only
    B. B only
    C. Both A and B
    D. Neither A nor B

35. Technician A says restricted exhaust may cause reduced fuel economy. Technician B says restricted exhaust causes reduced engine efficiency. Who is correct?

    A. A only
    B. B only
    C. Both A and B
    D. Neither A nor B

36. A hesitation during acceleration from a stop on a fuel-injected engine may be caused by all of the following EXCEPT:

    A. A purge-control solenoid
    B. A faulty manifold absolute-pressure sensor
    C. A faulty throttle-position sensor
    D. A faulty vehicle-speed sensor

37. An engine has a lack of power and excessive fuel consumption. Technician A says a broken timing belt could be the cause. Technician B says the timing belt may have jumped a tooth 180 degrees out of time. Who is correct?

    A. A only
    B. B only
    C. Both A and B
    D. Neither A nor B

38. What is the purpose of coating the back of an ignition module with heat sink grease (sometimes referred to as dielectric silicone grease) before installation?

    A. To insulate the module from excessive voltage spikes
    B. To electrically insulate the module from ground
    C. To ensure electrical ground
    D. To help dissipate heat from the module

39. Technician A says a damaged or missing air filter can increase wear on cylinder walls. Technician B says an air filter problem can affect fuel consumption. Who is correct?

    A. A only
    B. B only
    C. Both A and B
    D. Neither A nor B

40. An evaporative system DTC may be set by all of the following EXCEPT:

    A. A cracked vacuum hose
    B. A loose gas cap
    C. An open purge control solenoid
    D. A leaking fuel injector

41. Technician A says the PCM cannot be harmed with static electricity if the negative and positive battery cables are disconnected. Technician B says you should always ground yourself to the vehicle while working on a PCM. Who is correct?

    A. A only
    B. B only
    C. Both A and B
    D. Neither A nor B

42. All of the following are part of the diagnostic process EXCEPT:

    A. Verify the complaint
    B. Road test the vehicle
    C. Perform a visual inspection
    D. Re-flash the PCM

43. A sequential fuel-injected vehicle has a rough idle. Technician A says this could be caused by a cracked hose between the fuel tank and the EVAP canister. Technician B says a malfunctioning EVAP purge solenoid can cause idle problems. Who is correct?

    A. A only
    B. B only
    C. Both A and B
    D. Neither A nor B

44. While scanning an OBD II vehicle for DTCs, a P1000 is retrieved. Technician A says that a first digit of P means the code is a powertrain trouble code. Technician B says that a second digit of 1 means the code is a manufacturer-specific code. Who is correct?

    A. A only
    B. B only
    C. Both A and B
    D. Neither A nor B

45. Technician A says that worn valve train components usually produce a clicking noise. Technician B says an engine noise diagnosis should be performed before doing engine repair work. Who is correct?

    A. A only
    B. B only
    C. Both A and B
    D. Neither A nor B

46. Which of the following is the most likely symptom resulting from an evaporative emissions system failure?

    A. Increased tail pipe emissions
    B. Fuel odor
    C. Engine miss at highway speed
    D. Hard to start on a cold engine

47. A vehicle needs to have its PCM re-flashed. Technician A says the vehicle should be left running to prevent accidental battery discharge during flashing. Technician B says the scan tool must fit in the DLC snuggly. Who is correct?

    A. A only
    B. B only
    C. Both A and B
    D. Neither A nor B

48. When diagnosing a fuel injection system problem, a technical service bulletin search is performed for all of the following reasons EXCEPT:

    A. Year, make, and vehicle identification number
    B. Midyear production changes
    C. Service manual updated
    D. View revisions to existing procedures

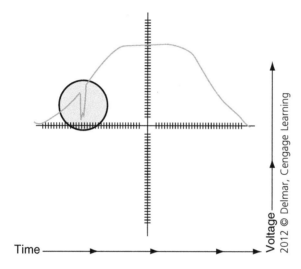

49. Technician A says the above illustration shows a defective throttle-position sensor. Technician B says it is a scope pattern of the manifold absolute-pressure sensor. Who is correct?

    A. A only
    B. B only
    C. Both A and B
    D. Neither A nor B

50. Technician A says a defective MAP sensor may cause a lean air/fuel ratio. Technician B says the MAP sensor can cause a no start. Who is correct?

    A. A only
    B. B only
    C. Both A and B
    D. Neither A nor B

PREPARATION EXAM 3

1. All of the following are tests performed on the PCV system EXCEPT:

   A. The rattle test

   B. The snap-back test

   C. Crankcase vacuum test

   D. Blowby test

2. Which of the following is the LEAST LIKELY cause of ignition module failure?

   A. An open spark plug wire

   B. No dielectric grease under the module

   C. A fouled spark plug

   D. Loose module mounting screws

3. A sequential fuel-injected vehicle has poor fuel economy, yet starts and runs fine. Technician A says the fuel return line may be partially restricted. Technician B says the fuel pressure regulator may be stuck open. Who is correct?

   A. A only

   B. B only

   C. Both A and B

   D. Neither A nor B

4. Technician A says if the passages in the exhaust gas recirculation (EGR) system get plugged up with carbon, then the engine could spark knock. Technician B says if the EGR passages get plugged up, then the engine will not run hot enough for complete combustion. Who is correct?

   A. A only

   B. B only

   C. Both A and B

   D. Neither A nor B

5. Technician A says a coil with weak reserve voltage could cause a miss under acceleration. Technician B says the ignition coil only produces enough secondary voltage to jump the spark plug gap. Who is correct?

   A. A only

   B. B only

   C. Both A and B

   D. Neither A nor B

6. Which of the following is the very first step a technician should take when performing a diagnostic procedure?

   A. Perform simple tests

   B. Retrieve diagnostic trouble codes

   C. Check for technical service bulletins

   D. Verify the customer complaint

7. All of the following could cause a multiple-cylinder misfire code EXCEPT:

    A. Low fuel pressure

    B. A leaking exhaust gas recirculation valve

    C. A burnt valve

    D. Retarded valve timing

8. A vehicle is equipped with a vented gas cap. Technician A says that if a non-vented cap is installed, then the vehicle could run rich at high speeds. Technician B says if a non-vented cap is installed in this vehicle, then the gas tank could collapse. Who is correct?

    A. A only

    B. B only

    C. Both A and B

    D. Neither A nor B

9. Technician A says secondary air injection is added to some vehicles to help oxidize hydrocarbon and carbon monoxide emissions. Technician B says secondary air is needed to lean out the air/fuel mixture on some vehicles. Who is correct?

    A. A only

    B. B only

    C. Both A and B

    D. Neither A nor B

10. Technician A says a scan tool is required to retrieve trouble codes on a vehicle with On-Board Diagnostics Second Generation (OBD II). Technician B says OBD II-compliant vehicles use a standardized trouble code format. Who is correct?

    A. A only

    B. B only

    C. Both A and B

    D. Neither A nor B

11. A Technician is performing a running compression test on a vehicle with suspected worn ring and cylinder problems. Technician A says running compression should be half of static compression at idle. Technician B says during a running compression test, the technician should increase the engine speed to 2,000 rpm, and the running compression should be lower than at idle. Who is correct?

    A. A only

    B. B only

    C. Both A and B

    D. Neither A nor B

12. Which of the following is LEAST LIKELY to be a test that would be performed on an ignition coil?

    A. Resistance primary to primary

    B. Resistance primary to secondary

    C. Secondary circuit amperage

    D. Primary circuit amperage

13. A vehicle with low engine power is being diagnosed. Technician A says an exhaust back-pressure test should be at least 5 psi at 2,000 rpm. Technician B says the catalytic converter may have come apart and be restricting exhaust flow. Who is correct?

    A. A only
    B. B only
    C. Both A and B
    D. Neither A nor B

14. Technician A says the filter on the engine-off natural vacuum pump (EONV) should be serviced with the regular oil change interval. Technician B says the EONV EVAP system does not use a vent valve. Who is correct?

    A. A only
    B. B only
    C. Both A and B
    D. Neither A nor B

15. All of the following apply to OBD II vehicles EXCEPT:

    A. A standardized 26-pin data link connector (DLC)
    B. A standardized list of DTCs
    C. The ability to perform rationality tests on components
    D. A standardized communication protocol

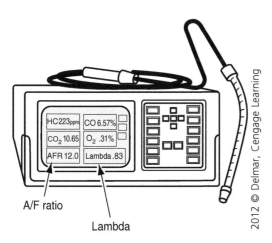

A/F ratio

Lambda

2012 © Delmar, Cengage Learning

16. Refer to the gas analyzer readings illustration. Technician A says the vehicle is running too rich. Technician B says the high hydrocarbon (HC) reading is from incomplete combustion. Who is correct?

    A. A only
    B. B only
    C. Both A and B
    D. Neither A nor B

17. A test light is connected between the negative side of the ignition coil and ground and the engine is cranked. Technician A says a flickering test light could be caused by a defective ignition module. Technician B says a flickering test light could be caused by a defective pickup coil. Who is correct?

    A. A only

    B. B only

    C. Both A and B

    D. Neither A nor B

18. A multi-port fuel injection vehicle has a tip-in hesitation when warm. All of these could be the cause EXCEPT:

    A. Throttle position sensor (TPS)

    B. Canister purge control solenoid

    C. EGR

    D. Defective secondary air pump motor

19. Technician A says that one of the first steps in diagnosing any EGR valve related concern is to check the vacuum and electrical connections. Technician B says that in many systems, the powertrain control module (PCM) uses other sensor inputs that could cause an EGR problem, and therefore DTCs should be corrected before replacing any EGR components. Who is correct?

    A. A only

    B. B only

    C. Both A and B

    D. Neither A nor B

OXYGEN SENSOR VOLTAGE VARIATIONS

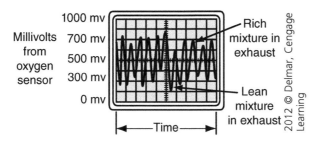

20. Based on the O_2 sensor wave form shown in the illustration, which of the following is true?

    A. This represents a lean-biased condition.

    B. The O_2 sensor is functioning correctly.

    C. This represents a lazy oxygen sensor.

    D. This represents a rich-biased condition.

21. Technician A says secondary air-injection systems must be monitored for proper operation on a vehicle certified as OBD II-compliant if equipped. Technician B says secondary air-injection systems are not monitored on OBD I-compliant vehicles. Who is correct?

    A. A only
    B. B only
    C. Both A and B
    D. Neither A nor B

22. A Hall-effect sensor is being tested. Technician A says the Hall-effect sensor should have a resistance value of 500-1,500 ohms. Technician B says a brass feeler gauge should be used to adjust a Hall-effect sensor. Who is correct?

    A. A only
    B. B only
    C. Both A and B
    D. Neither A nor B

23. All of the following are possible causes of turbocharger failure EXCEPT:

    A. Poor intake air filtration
    B. Poor engine oil maintenance
    C. Poor engine cooling system maintenance
    D. Poor exhaust system maintenance.

24. Technician A says the EVAP system assists in the reduction of oxides of nitrogen (NOx). Technician B says the EVAP system prevents fuel vapors from escaping into the atmosphere. Who is correct?

    A. A only
    B. B only
    C. Both A and B
    D. Neither A nor B

25. When testing for voltage drop in the power and ground distribution circuits, Technician A says the circuit being tested must be operating. Technician B says any corrosion adds unwanted resistance. Who is correct?

    A. A only
    B. B only
    C. Both A and B
    D. Neither A nor B

26. When the alternator belt and belt tension are satisfactory and the alternator output is low, Technician A says the alternator may be defective. Technician B says the problem could be high resistance in the alternator field circuit. Who is correct?

    A. A only
    B. B only
    C. Both A and B
    D. Neither A nor B

27. A pickup coil resistance is being tested with an ohmmeter. Technician A says when the pickup coil leads are moved, an erratic ohmmeter reading indicates the need for replacement. Technician B says that an infinite ohmmeter reading between the pickup coil terminals is normal on some pickup coils. Who is correct?

    A. A only
    B. B only
    C. Both A and B
    D. Neither A nor B

28. While testing fuel pressure on a multi-port fuel injection engine, Technician A says there will always be a Schrader test port for fuel system testing. Technician B says that a high fuel pressure reading could be the result of a plugged return line. Who is correct?

    A. A only
    B. B only
    C. Both A and B
    D. Neither A nor B

29. A stuck air-switching valve that constantly sends air pump output to the exhaust manifold will most likely result in which of the following conditions?

    A. Poor fuel economy
    B. A constant rich oxygen sensor signal
    C. Engine spark knock
    D. The exhaust to overheat

30. Technician A says some PCMs require a program chip to be transferred from the old PCM to the new PCM when replacing. Technician B says some PCMs require flash programming when replacing. Who is correct?

    A. A only
    B. B only
    C. Both A and B
    D. Neither A nor B

31. A vehicle emits a belt squeal when starting and on acceleration. Technician A says the alternator bearings may be defective. Technician B says the alternator belt automatic tensioner may be defective. Who is correct?

    A. A only
    B. B only
    C. Both A and B
    D. Neither A nor B

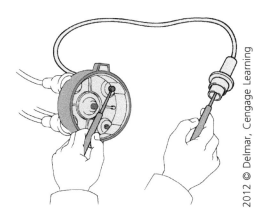

2012 © Delmar, Cengage Learning

32. Refer to the above illustration. Technician A says secondary circuit resistance is being checked. Technician B says high resistance in the shown component could lead to ignition coil failure. Who is correct?

   A. A only
   B. B only
   C. Both A and B
   D. Neither A nor B

33. When a vacuum leak is suspected for a high idle complaint, the preferred testing method would be:

   A. Propane
   B. Smoke
   C. Water
   D. Throttle body cleaner

34. Technician A says the PCM controls the amount of canister purge based on other inputs. Technician B says the EVAP system controls the amount of carbon monoxide (CO) emissions produced by the engine. Who is correct?

   A. A only
   B. B only
   C. Both A and B
   D. Neither A nor B

35. A TPS is being tested with a digital multimeter (DMM). Technician A says the voltage on the signal wire should be around 1.0 volt or less at idle. Technician B says the voltage reading on the reference wire should be around 5 volts. Who is correct?

   A. A only
   B. B only
   C. Both A and B
   D. Neither A nor B

36. Which of the following conditions is LEAST LIKELY to be diagnosed with an emissions analyzer?

    A.   Cylinder misfire
    B.   Cylinder efficiency
    C.   Head gasket integrity
    D.   Exhaust system leakage

37. Technician A says an injector with a lower pressure drop during an injector pressure-drop test indicates a rich-running injector. Technician B says an injector with a lower pressure drop could be leaking. Who is correct?

    A.   A only
    B.   B only
    C.   Both A and B
    D.   Neither A nor B

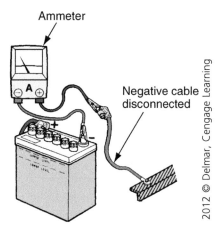

Ammeter

Negative cable disconnected

2012 © Delmar, Cengage Learning

38. A DMM set in the milliamp position is connected in series between the negative battery terminal and the negative cable, as shown. What is being measured?

    A.   Voltage drop
    B.   Open circuit voltage
    C.   Parasitic drain
    D.   Charging system amperage

39. When replacing a PROM, Technician A says that you should always ground yourself to the vehicle. Technician B says that the PCM connector should be disconnected with the key on to prevent static charge. Who is correct?

    A.   A only
    B.   B only
    C.   Both A and B
    D.   Neither A nor B

40. Technician A says the throttle body must be completely disassembled before soaking it in cleaning solvent. Technician B says some throttle bodies cannot be cleaned in solvent. Who is correct?

    A. A only
    B. B only
    C. Both A and B
    D. Neither A nor B

41. Technician A says that an overfilled crankcase can cause hydraulic lifter noise due to oil aeration. Technician B says using motor oil with a viscosity rating that is too low can cause hydraulic lifter noise. Who is correct?

    A. A only
    B. B only
    C. Both A and B
    D. Neither A nor B

42. Technician A says that a DMM can be used to check an oxygen sensor. Technician B says that to check an oxygen sensor, you can use an oscilloscope. Who is correct?

    A. A only
    B. B only
    C. Both A and B
    D. Neither A nor B

43. When testing the engine coolant temperature sensor circuit, all of the following are used EXCEPT:

    A. Scan tool data
    B. Resistance value
    C. Voltage value
    D. Amperage value

44. A vehicle is being diagnosed for an overheating complaint while at idle. Technician A says the electric cooling fan may be defective. Technician B says that the thermostat may be opening too soon. Who is correct?

    A. A only
    B. B only
    C. Both A and B
    D. Neither A nor B

45. Resistance on a negative-temperature coefficient coolant sensor is being tested against specifications. Technician A says a resistance reading lower than specifications would send a signal indicating a warmer-than-actual engine temperature. Technician B says if the resistance is lower than specifications, the engine may exhibit hard starting when warm. Who is correct?

    A. A only
    B. B only
    C. Both A and B
    D. Neither A nor B

46. A vacuum gauge indicates low vacuum (12 inches of mercury). Technician A says late valve timing will cause a low vacuum reading. Technician B says to connect the gauge to a venturi vacuum port. Who is correct?

    A. A only
    B. B only
    C. Both A and B
    D. Neither A nor B

47. Technician A says the PCM will increase the fuel injector pulse width if there is no oxygen in the exhaust. Technician B says if there is a lean condition, then the short-term fuel trim (SFT) will show a minus value on the scan tool. Who is correct?

    A. A only
    B. B only
    C. Both A and B
    D. Neither A nor B

48. Which of the follow is the LEAST LIKELY cause of alternator drive belt noise?

    A. Defective belt tensioner
    B. Glazed belt
    C. Belt alignment
    D. Over-tighten belt

49. Which of the follow diagnostic trouble codes are Type A codes and should be troubleshot first?

    A. Transmission
    B. Vehicle speed control
    C. Misfire control
    D. Non-emissions-related

50. What is the maximum voltage drop for a PCM ground circuit?

    A. 0.10 volts
    B. 0.30 volts
    C. 0.20 volts
    D. 1.0 volts

PREPARATION EXAM 4

1. Technician A says if the PCV valve is stuck closed, then excessive crankcase pressure forces blowy gases through the clean air hose into the air filter. Technician B says if the positive crankcase ventilation (PCV) valve is stuck open, then excessive airflow through the valve causes a rich air/fuel ratio. Who is correct?

    A. A only
    B. B only
    C. Both A and B
    D. Neither A nor B

2. Technician A says that a 12-volt test light connected between the negative side of an ignition coil and ground that blinks on and off during cranking confirms the primary circuit is being switched. Technician B says any voltage drops greater than .3 volts in the primary circuit can reduce secondary circuit kV output. Who is correct?

   A. A only
   B. B only
   C. Both A and B
   D. Neither A nor B

3. A 6-cylinder engine is making a loud metallic knocking that gets louder as the engine warms up or if the throttle is quickly snapped open. The noise almost disappears when the spark for cylinder #2 is shorted to ground. Technician A says the problem could be a cracked flex plate. Technician B says the problem is most likely a loose connecting rod bearing. Who is correct?

   A. A only
   B. B only
   C. Both A and B
   D. Neither A nor B

4. All of the following symptoms can be caused by low fuel pressure EXCEPT:

   A. Engine surge
   B. Strong sulfur smell from the exhaust
   C. Low engine power
   D. Limited top speed

5. All of the following statements about powertrain control module (PCM) inputs are true EXCEPT:

   A. High-impedance digital volt/ohm meters may be used for diagnosis.
   B. The O_2 sensor produces very low voltage.
   C. An analog meter may be used for diagnosis.
   D. Most inputs use a 5-volt reference voltage.

6. A technician is performing a compression test. Which statement below is most likely true?

   A. A low reading on one cylinder may be caused by a vacuum leak at that cylinder.
   B. All cylinders reading even, but lower than normal, may be caused by a slipped timing chain.
   C. Low readings on two adjacent cylinders may be caused by carbon buildup.
   D. All cylinders with higher than normal readings could be caused by a blown head gasket.

7. Technician A says the thermostatic coil controls the opening and closing of the orifice inside the coupling on a viscous fan clutch. Technician B says when the thermostatic coil is cold, the orifice is open on a viscous fan clutch. Who is correct?

   A. A only
   B. B only
   C. Both A and B
   D. Neither A nor B

8. Technician A says when testing an ignition related no-start problem, the technician should always check for available spark at an ignition wire first. Technician B says if a test light is connected to the negative side of the coil while cranking and the test light flickers, then the secondary ignition system needs to be tested. Who is correct?

   A. A only

   B. B only

   C. Both A and B

   D. Neither A nor B

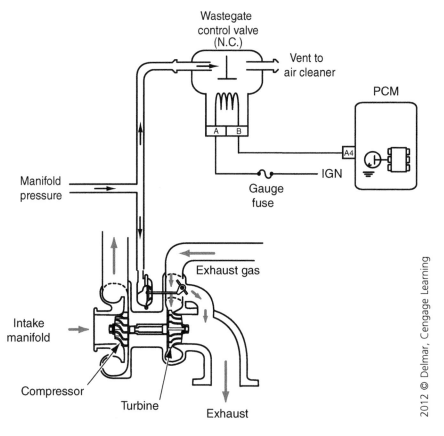

9. Refer to the above illustration. All of the following conditions could reduce engine power during a boost condition EXCEPT:

   A. Damaged compressor fins

   B. Wastegate valve stuck open

   C. A blown gauge fuse

   D. A leak between the turbocharger and exhaust manifold

10. Technician A says that when vacuum is applied to some exhaust gas recirculation (EGR) valves with the engine idling, the EGR valve should open and idle should become erratic. Technician B says that some EGR valves can be opened with a scan tool. Who is correct?

    A. A only

    B. B only

    C. Both A and B

    D. Neither A nor B

11. Technician A says to use a scan tool to verify throttle position sensor input and related diagnostic trouble codes (DTCs). Technician B says coolant-temperature sensor input is used to help determine open- and closed-loop status. Who is correct?

    A. A only
    B. B only
    C. Both A and B
    D. Neither A nor B

12. Technician A says a scan tool can be used to check for oxygen sensor codes and operation. Technician B says. The oxygen sensor can be removed and tested under for proper operation. Who is correct?

    A. A only
    B. B only
    C. Both A and B
    D. Neither A nor B

13. Technician A says that an oscilloscope can be used to watch the mass air flow (MAF) sensor signal switch from idle to wide-open throttle (WOT) status. Technician B says fuel injectors can only be tested using an ohmmeter. Who is correct?

    A. A only
    B. B only
    C. Both A and B
    D. Neither A nor B

14. While performing a valve adjustment, Technician A says the crankshaft must be placed in position so the piston is at top dead center (TDC) on the exhaust stroke. Technician B says that adjusting valves with too much clearance may cause rough running and burnt valves. Who is correct?

    A. A only
    B. B only
    C. Both A and B
    D. Neither A nor B

15. Technician A says an ignition coil should be tested for both primary and secondary winding resistance. Technician B says available coil output can be tested with an oscilloscope. Who is correct?

    A. A only
    B. B only
    C. Both A and B
    D. Neither A nor B

16. Technician A says if an engine uses low-resistance injectors with less than 3 ohms resistance, then the PCM will use a current-limiting or peak and hold injector driver to operate the fuel injector. Technician B says using a current-limiting or peak and hold injector increases fuel injector noise. Who is correct?

    A. A only
    B. B only
    C. Both A and B
    D. Neither A nor B

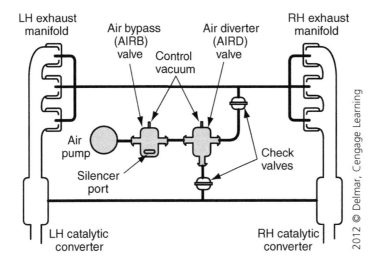

17. Refer to the above illustration. Technician A says the check valves allow air in the exhaust without exhaust getting in the air diverter valve. Technician B says the air bypass valve prevents secondary air from entering the exhaust on deceleration. Who is correct?

    A. A only
    B. B only
    C. Both A and B
    D. Neither A nor B

18. Technician A says a manifold absolute pressure (MAP) sensor should be able to hold vacuum during a test. Technician B says some MAP sensors produce an analog voltage signal while others produce a digital square wave signal. Who is correct?

    A. A only
    B. B only
    C. Both A and B
    D. Neither A nor B

19. The result of a battery load test done at 78°F with a carbon pile load tester is 8.9 volts. Technician A says minimum load test voltage is 9.6 volts, and this result is unacceptable. Technician B says that you should compare the results to the tables from the tool or battery manufacturer for temperature correction. Who is correct?

    A. A only
    B. B only
    C. Both A and B
    D. Neither A nor B

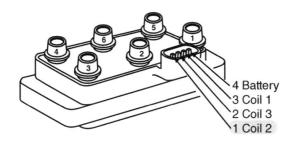

4 Battery
3 Coil 1
2 Coil 3
1 Coil 2

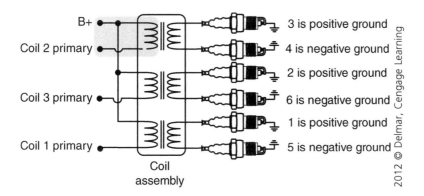

20. Technician A says the schematic in the above illustration is for a COP ignition system. Technician B says if coil primary #2 became open, then two cylinders would be killed. Who is correct?

    A. A only
    B. B only
    C. Both A and B
    D. Neither A nor B

21. Which of the following is the LEAST LIKELY cause of poor fuel mileage on a vehicle with sequential fuel injection (SFI)?

    A. Restricted pressure regulator
    B. A defective oxygen sensor
    C. Restricted exhaust
    D. Defective speed sensor

22. All of the following could cause an EVAP system large leak code to be set EXCEPT:

    A. Loose gas cap
    B. Leaking injector O-ring
    C. Cracked purge control solenoid
    D. Cracked canister vent hose

23. Technician A says that for non-emissions-related DTCs, you can replace the component without using the flow chart. Technician B says depending on the DTC set, some steps of the flow chart can be bypassed. Who is correct?

    A. A only
    B. B only
    C. Both A and B
    D. Neither A nor B

24. Low battery or system voltage can cause all of the following EXCEPT:

   A. A code to be set

   B. Increased idle RPM

   C. Increased steering effort

   D. Poor drivability at high speeds

25. All of the following could cause a cylinder misfire diagnostic code to be set EXCEPT:

   A. Low fuel pump pressure

   B. Secondary insulation breakdown

   C. Worn cam shaft lobe

   D. Erratic crank shaft sensor signal

26. Technician A says that nylon fuel line should be routed in a way to prevent kinking. Technician B says nylon fuel line can be repaired on some vehicles; if not, the entire line must be replaced. Who is correct?

   A. A only

   B. B only

   C. Both A and B

   D. Neither A nor B

27. Technician A says that when vacuum is applied to the EGR valve with the engine idling, the EGR valve should open and idle should increase. Technician B says that a diagnosis of the EGR valve should be done with the engine idling. Who is correct?

   A. A only

   B. B only

   C. Both A and B

   D. Neither A nor B

28. Technician A says DTC should be erased using a scan tool. Technician B says the monitor for the related fault should be run before the vehicle is returned to the customer. Who is correct?

   A. A only

   B. B only

   C. Both A and B

   D. Neither A nor B

29. Technician A says to repeat the pressure test after repairs are made to the cooling system to ensure that all leaks are found. Technician B says a cooling system pressure test should include testing the radiator cap. Who is correct?

   A. A only

   B. B only

   C. Both A and B

   D. Neither A nor B

30. The PCM sends a digital signal to the ignition control module to control which of the following?

    A. RPM
    B. Ignition timing
    C. Cylinder identification
    D. EGR

31. Technician A says a fuel-pressure test is performed to test fuel pump operation. Technician B says it is possible to have a good fuel pressure reading and insufficient fuel flow. Who is correct?

    A. A only
    B. B only
    C. Both A and B
    D. Neither A nor B

32. Technician A says charcoal canister filters are no longer serviceable. Technician B says gas caps with pressure and vacuum valves must be checked for leakage with a pressure tester. Who is correct?

    A. A only
    B. B only
    C. Both A and B
    D. Neither A nor B

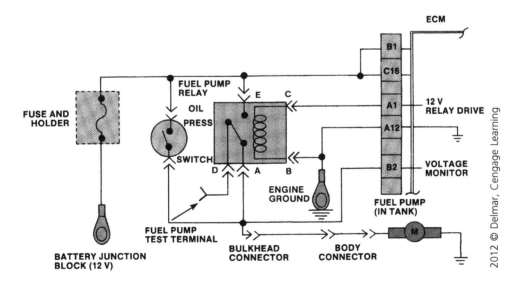

33. A vehicle with the fuel system shown in the above illustration is being diagnosed for a long cranking before start complaint. Technician A says the fuse could be blown, causing the long crank time before staring complaint. Technician B says this system type uses the oil pressure circuit as a back-up to the fuel pump relay circuit. Who is correct?

    A. A only
    B. B only
    C. Both A and B
    D. Neither A nor B

34. All of the following conditions could result in a voltage drop in a circuit EXCEPT:

    A. Spread terminals in a connector
    B. Broken strands in a stranded wire
    C. Greenish corrosion builds up in a connector
    D. Using dielectric grease in a connector

35. A vehicle is being diagnosed for a poor drivability complaint. The vehicle also fails to start at times; it just cranks with no start. Technician A says the charging system could be undercharging at times. Technician B says an under- or overcharging system should generate a trouble code. Who is correct?

    A. A only
    B. B only
    C. Both A and B
    D. Neither A nor B

36. All of the following are true when testing a pickup coil EXCEPT:

    A. Resistance value should fall within manufacturer's specifications.
    B. Pickup coils should produce an AC voltage signal.
    C. If adjustable, then the gap should be checked with a steel feeler gauge.
    D. An erratic ohmmeter reading while wiggling the pickup coil wires indicates that the pickup coil has an intermittent open.

37. Nylon fuel lines should be inspected for all of the following EXCEPT:

    A. Discoloration
    B. Loose fitting
    C. Cracks
    D. Kinks

38. Technician A says that when checking a pulsed-air secondary air-injection system, exhaust pressure pulses felt at the fresh-air intake hose indicate a bad check or reed valve. Technician B says that when testing secondary air-injection systems, the gas analyzer can be used to confirm normal air-injection operation. Who is correct?

    A. A only
    B. B only
    C. Both A and B
    D. Neither A nor B

39. Technician A says a digital voltmeter cannot be used to check an O_2 sensor. Technician B says a test light can be used to check an O_2 sensor. Who is correct?

    A. A only
    B. B only
    C. Both A and B
    D. Neither A nor B

40. All of the following are measured by a 5-gas analyzer EXCEPT:

    A. Oxides of nitrogen

    B. Smog

    C. Carbon monoxide

    D. Hydrocarbons

41. Technician A says the alternator connectors should always be inspected for corrosion and/or distortion from overheating when replacing an alternator. Technician B says some replacement alternators come with a new connector. Who is correct?

    A. A only

    B. B only

    C. Both A and B

    D. Neither A nor B

42. Technician A says most 2-wire COP systems use individual ignition control modules for each cylinder. Technician B says a 4-wire COP coil is checked like any other ignition coil. Who is correct?

    A. A only

    B. B only

    C. Both A and B

    D. Neither A nor B

43. An SFI vehicle stalls intermittently at idle and has negative long-term fuel trim correction values stored when checked with a scan tool. All of the following conditions could cause this EXCEPT:

    A. Leaking fuel injectors

    B. A leaking fuel pressure regulator diaphragm

    C. A vacuum leak

    D. A fuel pressure regulator stuck closed

44. Technician A says a stuck-open purge control valve could cause a hesitation on takeoff. Technician B says the purge control valve is a normally open solenoid. Who is correct?

    A. A only

    B. B only

    C. Both A and B

    D. Neither A nor B

45. An SFI engine has poor acceleration when the vehicle is suddenly accelerated to wide-open throttle. Idle and cruise performance is fine. Technician A says a faulty mass airflow sensor could cause this. Technician B says a weak fuel pump could cause this. Who is correct?

    A. A only

    B. B only

    C. Both A and B

    D. Neither A nor B

46. All of the following are part of reprogramming the PCM EXCEPT:

    A. Connect a special type of battery charger to vehicle

    B. Connect scan tool to diagnostic link connector (DLC)

    C. Start the engine for 15 seconds

    D. Validate the vehicle identification number (VIN)

47. During a vacuum test the vacuum gauge shows a rapidly fluctuating motion from 15 to 21 in. Hg at idle. Technician A says this could be caused by a loose intake manifold. Technician B says this could be caused by a burned exhaust valve. Who is correct?

    A. A only

    B. B only

    C. Both A and B

    D. Neither A nor B

48. Technician A says a faulty throttle position sensor (TPS) can cause a dead spot in the throttle when accelerating. Technician B says faulty TPS will always set a DTC. Who is correct?

    A. A only

    B. B only

    C. Both A and B

    D. Neither A nor B

49. During a cylinder leak down test on a 6-cylinder engine, air is heard coming from the #2 spark plug hole as cylinder #3 is being checked. Technician A says that this could be caused by a blown head gasket. Technician B says this could be caused by a cracked engine block. Who is correct?

    A. A only

    B. B only

    C. Both A and B

    D. Neither A nor B

50. Technician A says when high voltage drop is found in a circuit, check for burned wires, connector ring terminals, loose retaining nuts, or other wire and connector concerns. Technician B says that greenish white corrosion can happen at any point where the wire insulation has been pierced or opened in any way. Who is correct?

    A. A only

    B. B only

    C. Both A and B

    D. Neither A nor B

PREPARATION EXAM 5

1. With a thermometer taped to the upper radiator hose and the vehicle fully warmed up, the temperature indication should be which of the following?

    A. Lower than the temperature measured at the lower radiator hose
    B. One-third of the engine thermostat temperature rating
    C. Ambient temperature
    D. Within a few degrees of the engine thermostat temperature rating

2. Technician A says timing specifications can be found on the under hood vehicle emissions control information sticker (VECI). Technician B says the late ignition timing cause the engine to crank slow on a warm engine. Who is correct?

    A. A only
    B. B only
    C. Both A and B
    D. Neither A nor B

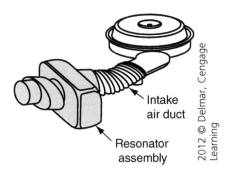

Intake air duct

Resonator assembly

2012 © Delmar, Cengage Learning

3. Refer to the above illustration. Technician A says the resonator assembly is used to reduce noise. Technician B says a leak in the intake air duct would allow unfiltered air to enter the engine. Who is correct?

    A. A only
    B. B only
    C. Both A and B
    D. Neither A nor B

4. All of the following are Components of a PCV system EXCEPT:

    A. PCV heater
    B. Valve grommet
    C. Inlet filter
    D. Purge control solenoid

5. Technician A says continuous monitors consist of catalyst efficiency, misfire, and component monitors. Technician B says the continuous monitors consist of three separate monitors. Who is correct?

    A. A only
    B. B only
    C. Both A and B
    D. Neither A nor B

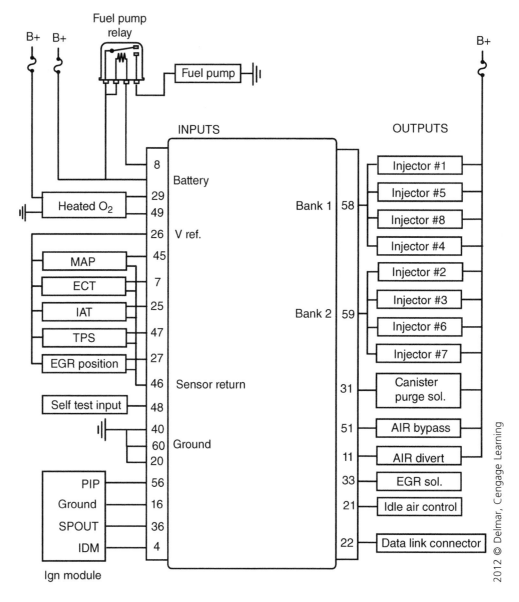

6. Refer to the above illustration. A multi-port, fuel-injected distributor-type ignition V8 engine is running rough and has a lean fuel mixture. The fuel injectors are pair-fired in two groups consisting of cylinders #2, #3, #6 and #7 and cylinders #1, #4, #5 and #8. Injecting propane will smooth out engine idle quality. There are no vacuum leaks. After performing a cylinder balance test, cylinders #2, #3 and #6 are found to be weak. Technician A says a bad ignition module could cause this. Technician B says the low resistance on cylinder #2 fuel injector could cause this. Who is correct?

   A.   A only

   B.   B only

   C.   Both A and B

   D.   Neither A nor B

7. Technical service bulletins should be used for all of the following EXCEPT:

   A. Updated service manual procedures

   B. Updated repair procedures for certain complaints

   C. Updated vehicle recalls

   D. Updated flash programming

8. All of the following are inputs for electronic timing control EXCEPT:

   A. Knock sensor

   B. Throttle position sensor

   C. Engine coolant temperature

   D. Oxygen sensor

9. Technician A says that if some EGR passages are plugged in the intake manifold, then the engine may have detonation when accelerating. Technician B says that plugged EGR passages may cause a failed IM240 emissions test for high oxides of nitrogen. Who is correct?

   A. A only

   B. B only

   C. Both A and B

   D. Neither A nor B

10. All of the following types of test equipment are used to check the oxygen sensor EXCEPT:

   A. Lab scope

   B. Low-impedance volt meter

   C. Graphing meter

   D. High-impedance volt meter

11. During a cylinder power balance test, there is no RPM drop on cylinder #4. Technician A says that the cylinder is not contributing to the power flow of the engine. Technician B says that the cylinder may have an inoperative spark plug. Who is correct?

   A. A only

   B. B only

   C. Both A and B

   D. Neither A nor B

12. A vehicle with electronic spark timing control has low power and needs its ignition timing adjusted. Technician A says ignition timing on a distributor-type ignition is usually set with the engine running and the computer control circuit disconnected or disabled. Technician B says ignition timing is usually set with the engine running at 2,500 rpm. Who is correct?

   A. A only

   B. B only

   C. Both A and B

   D. Neither A nor B

13. A vehicle is being diagnosed for a PO134 trouble code. (Oxygen Sensor Circuit, No Activity Detected) (bank #1, sensor #1) with a bank #1 fuel trim of +20. Technician A says the problem could be high fuel system pressure. Technician B says the problem is more likely confined to the downstream oxygen sensor on the bank were cylinder #1 is on. Who is correct?

    A. A only
    B. B only
    C. Both A and B
    D. Neither A nor B

14. The pulse air-injection system is driven by which of the following?

    A. A drive belt
    B. A helical gear
    C. An electric motor
    D. Negative pressure pulses in the exhaust.

15. Technician A says a voltage drop test checks the amount of resistance between two test points in a circuit. Technician B says more than a 0.5 volts drop indicates excessive resistance across the battery positive cable. Who is correct?

    A. A only
    B. B only
    C. Both A and B
    D. Neither A nor B

16. A proper cooling system inspection involves all of the following EXCEPT:

    A. A thermostat operation inspection
    B. A coolant condition inspection
    C. A cooling fan operation inspection
    D. A condenser inspection

17. High carbon monoxide (CO) emissions may be caused by all the following EXCEPT:

    A. Rich air/fuel mixture
    B. Exhaust manifold leak
    C. Fouled spark plug
    D. Malfunctioning secondary air switching valve

18. A vehicle with an electronic ignition fails to start. Technician A says this could be caused by a defective cam shaft sensor connection. Technician B says this could be caused by an ignition coil. Who is correct?

    A. A only
    B. B only
    C. Both A and B
    D. Neither A nor B

19. Technician A says an enhanced EVAP system must be able to detect a leak as small as 0.020 inches. Technician B says the enhanced system must be able to detect a loose gas cap. Who is correct?

    A. A only
    B. B only
    C. Both A and B
    D. Neither A nor B

20. A PCM is being replaced. Technician A says the new PCM should be ordered using the original PCM's part number. Technician B says installing a used PCM can cause the theft deterrent system to activate on some vehicles. Who is correct?

    A. A only
    B. B only
    C. Both A and B
    D. Neither A nor B

21. While testing a turbocharger, the maximum boost pressure observed is 4 psi (27.6 kPa), while the specified pressure is 9 psi (62 kPa). Technician A says the engine compression may be low. Technician B says the wastegate may be sticking closed. Who is correct?

    A. A only
    B. B only
    C. Both A and B
    D. Neither A nor B

22. A vehicle has had its PCM replaced for a charging system problem; now the vehicle will not start. Technician A says the replacement PCM on some vehicles require being flash-programmed in order to operate. Technician B says the flash program for some scan tools can be downloaded and installed after the PCM is in the vehicle. Who is correct?

    A. A only
    B. B only
    C. Both A and B
    D. Neither A nor B

23. A timing belt that has jumped one tooth can cause all EXCEPT:

    A. Low power
    B. Poor fuel economy
    C. High engine idle
    D. Poor vehicle stopping

24. Which of the following conditions would most likely cause weak spark at all the spark plug wires?

    A. High primary circuit resistance
    B. Low secondary resistance
    C. Secondary spark plug wire insulation breakdown
    D. Ignition timing out of adjustment

25. Technician A says a leaking injector could cause a high CO_2 reading at idle. Technician B says that no vacuum to the pressure regulator could cause a high CO reading at idle. Who is correct?

    A. A only
    B. B only
    C. Both A and B
    D. Neither A nor B

26. Technician A says if the EGR valve remains open at idle and low speed, then the idle will be high. Technician B says if the EGR valve does not open at cruising speeds, then detonation can occur. Who is correct?

    A. A only
    B. B only
    C. Both A and B
    D. Neither A nor B

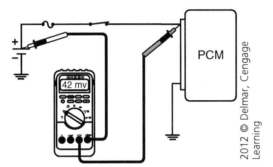

27. Technician A says that when measuring voltage drop as shown in the illustration, it must be done with a digital multimeter and the circuit that is being tested must be operating. Technician B says a high voltage drop reading indicates low resistance in the circuit. Who is correct?

    A. A only
    B. B only
    C. Both A and B
    D. Neither A nor B

28. Technician A says an OBD II warm-up cycle is defined as a trip in which the engine temperature increases by at least 40°F (22°C) and reaches 160°F (70°C) during one key on and the engine running (KOER) cycle. Technician B says a drive cycle refers to a trip in which the operating parameters required for the PCM to run OBD II emissions-related monitors has been met. Who is correct?

    A. A only
    B. B only
    C. Both A and B
    D. Neither A nor B

29. A restricted exhaust will cause vacuum readings to do what?

    A. Be 2 to 3 inches higher than normal
    B. Be 2 to 3 inches lower than normal
    C. Gradually drop as engine speed is increased
    D. Fluctuate between 17 and 20 inches

30. A charging system that is overcharging can be caused by all of the following EXCEPT:

    A. An internally shorted battery

    B. Shorted field windings

    C. A defective voltage regulator

    D. High generator field winding circuit resistance

31. An engine equipped with a distributorless ignition system will not start. Technician A says a defective crankshaft position sensor could cause this. Technician B says a shorted ground wire to the distributorless ignition systems (DIS) assembly could be the cause. Who is correct?

    A. A only

    B. B only

    C. Both A and B

    D. Neither A nor B

32. A vehicle with a lean exhaust code is being diagnosed. Technician A says the fuel pressure should be checked. Technician B says remove the vacuum hose from the fuel pressure regulator and check for fuel at the regulator vacuum nipple. Who is correct?

    A. A only

    B. B only

    C. Both A and B

    D. Neither A nor B

33. Which valves are normally closed on an evaporative emissions control system?

    A. Canister purge valve

    B. Both canister purge and canister vent valve

    C. Canister vent valve

    D. Neither canister purge nor canister vent valve

34. The PCM will automatically clear a Type B DTC if there are no additional faults detected after which of the following?

    A. Eighty warm-up cycles

    B. Forty consecutive warm-up cycles

    C. Two consecutive trips

    D. Four key-on/key-off cycles

35. A vehicle with a SFI V6 engine and OBD II emissions controls has set a DTC PO172 (System Too Rich, Bank 1). No other drivability concerns are present. The freeze frame data shows the code was set under warm idle conditions. Technician A says the problem could be an intake manifold vacuum leak. Technician B says the problem could be a defective fuel pressure regulator. Who is correct?

    A. A only

    B. B only

    C. Both A and B

    D. Neither A nor B

36. A vacuum leak is the suspected cause for a rough idle concern. Using a 5-gas analyzer, Technician A says O_2 will be higher than normal if a vacuum leak is present. Technician B says CO will be higher than normal if a vacuum leak is present. Who is correct?

    A. A only
    B. B only
    C. Both A and B
    D. Neither A nor B

37. To check ignition coil available voltage output, the technician should do which of the following?

    A. Disconnect the plug wire at the plug and ground it
    B. Disconnect the fuel pump power lead
    C. Disconnect the coil wire and ground it
    D. Conduct the test using a suitable spark tester that requires 25 kV

38. A turbocharged engine is experiencing excessive oil consumption and blue smoke from the tailpipe at idle and cruising speeds. This could be caused by which of the following?

    A. Dirty air filter
    B. A plugged oil return passage
    C. Restricted exhaust
    D. Turbo spinning too fast

39. Technician A says if the catalytic converter is restricted, then the engine will produce higher than normal vacuum at 2,000 rpm. Technician B says if you tap on a monolithic (honeycomb) converter with a rubber hammer and the converter rattles, then it should be replaced. Who is correct?

    A. A only
    B. B only
    C. Both A and B
    D. Neither A nor B

40. Technician A says the engine coolant temperature sensor voltage drop decreases as the coolant temperature increases. Technician A says the engine coolant temperature sensor resistance increases as the coolant temperature increases. Who is correct?

    A. A only
    B. B only
    C. Both A and B
    D. Neither A nor B

41. Technician A says incorrect cam shaft timing can cause an engine not to start. Technician B says incorrect cam shaft timing may cause a power loss. Who is correct?

    A. A only
    B. B only
    C. Both A and B
    D. Neither A nor B

42. Technician A says a scan tool can be used in conjunction with a thermometer to check thermostat operation. Technician B says thermostat operation can also be checked visually by running the engine from a cold start until it is hot. Who is correct?

    A. A only
    B. B only
    C. Both A and B
    D. Neither A nor B

43. Spark plug wires are being measured for resistance. Technician A says if the ohmmeter reads OL, then the circuit has little or no resistance. Technician B says plug wires should measure about 5,000 to 10,000 ohms per foot. Who is correct?

    A. A only
    B. B only
    C. Both A and B
    D. Neither A nor B

44. Which of the following is the least likely cause of poor fuel mileage on a vehicle with SFI?

    A. Plugged vacuum hose to the fuel pressure regulator
    B. Plugged return fuel line
    C. Plugged fuel filter
    D. Defective thermostat

45. All of the following are part of the evaporative emissions system EXCEPT:

    A. Roll-over valve
    B. Vapor canister
    C. Gas cap
    D. Pulse air feeder

46. Technician A says many different adapters are needed if a scan tool is going to be used to retrieve DTCs from different manufactured ODB II vehicles. Technician B says many different adapters are needed if a scan tool is going to be used to retrieve diagnostic trouble codes from different manufactured ODB I vehicles. Who is correct?

    A. A only
    B. B only
    C. Both A and B
    D. Neither A nor B

47. A vehicle has excessively high hydrocarbon emissions with no diagnostic trouble codes. Technician A says weak cylinder compression could be the cause. Technician B says a vacuum leak could be the cause. Who is correct?

    A. A only

    B. B only

    C. Both A and B

    D. Neither A nor B

48. Air is escaping from the PCV valve opening in the valve cover during a cylinder leakage test. Technician A says a blown head gasket is a possible cause. Technician B says air escaping from the PCV valve opening in the valve cover is leaking past the rings. Who is correct?

    A. A only

    B. B only

    C. Both A and B

    D. Neither A nor B

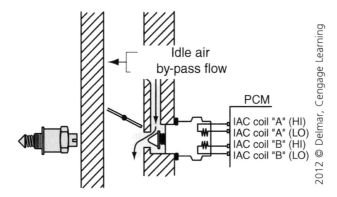

49. A vehicle with the illustrated idle control system has a high idle. Technician A says a break in any of the wires can cause a high idle depending on the position of the plunger at the time of the break. Technician B says coil A high and coil B high are for high idle, and coil A low and coil B low are for low idle. Who is correct?

    A. A only

    B. B only

    C. both A and B

    D. Neither A nor B

50. Technician A says a bi-directional scan tool is needed to actuate outputs such as the injectors for diagnostic purposes. Technician B says a bi-directional scan tool is needed to retrieve diagnostic trouble codes on an OBD II vehicle. Who is correct?

    A. A only
    B. B only
    C. Both A and B
    D. Neither A nor B

PREPARATION EXAM 6

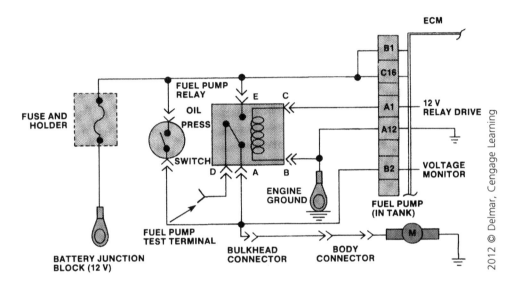

1. Refer to the above illustration. Technician A says the double arrow (>>) indicates a male and female connector. Technician B says if there is only one arrow (>), then the connector does not connect with another connector. Who is correct?

    A. A only
    B. B only
    C. Both A and B
    D. Neither A nor B

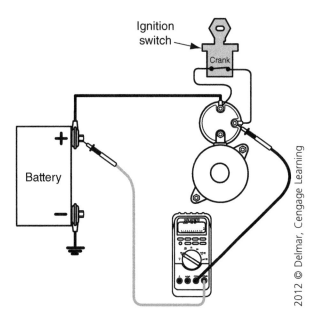

Ignition switch

Crank

Battery

2012 © Delmar, Cengage Learning

2. In the above illustration, a voltage drop test of the starter motor control circuit is being performed. With the ignition disabled, the engine is cranked over with the voltmeter on the lowest scale. Technician A says that when a reading above 1 volt is obtained, individual components in the circuit will need to be tested to find the problem. Technician B says that a reading below 3.5 volts is OK. Who is correct?

    A. A only
    B. B only
    C. Both A and B
    D. Neither A nor B

3. The ignition control module uses a digital signal from the PCM for what purpose?

    A. RPM signal
    B. Cylinder ID
    C. Timing control
    D. Synchronization signal

4. Technician A says a small amount of anti-seize should be used when replacing an oxygen sensor. Technician B says zirconium oxygen sensors require the use of thread locker when replacing. Who is correct?

    A. A only
    B. B only
    C. Both A and B
    D. Neither A nor B

5. The plunger of a PCV valve is stuck in the maximum flow position. Technician A says this can cause a rough idle. Technician B says this could cause excessive oil consumption. Who is correct?

    A. A only
    B. B only
    C. Both A and B
    D. Neither A nor B

6. A vehicle with a no-start condition in being diagnosed. The vehicle has secondary ignition voltage at the spark plug end, but no fuel pressure. The vehicle registers 12 volts at the fuel pump connector at key on and while cranking, but the fuel pump will not operate. Technician A says the fuel pump could be bad and might need to be replaced. Technician B says the positive and ground side of the fuel pump should be checked for voltage drop first. Who is correct?

   A. A only
   B. B only
   C. Both A and B
   D. Neither A nor B

7. Regardless of sensor type, what attribute of a speed sensor signal does the powertrain control module (PCM) process to determine the vehicle speed?

   A. Amplitude
   B. Voltage
   C. Frequency
   D. Pulse width.

8. The voltage signal of a negative temperature coefficient sensor will do what?

   A. Decrease as temperature goes up
   B. Increase as temperature goes up
   C. Be around 12 volts on most applications
   D. Be low if the sensor is open

9. Technician A says that a defective starter drive gear can cause a starter to have a whining noise. Technician B says a defective starter solenoid can cause a starter to whine. Who is correct?

   A. A only
   B. B only
   C. Both A and B
   D. Neither A nor B

10. A vehicle with a manual transmission will not crank. Technician A says the neutral safety switch might be stuck open. Technician B says the starter solenoid might be stuck closed. Who is correct?

   A. A only
   B. B only
   C. Both A and B
   D. Neither A nor B

11. An ammeter is used on the ignition primary circuit to check what?

   A. Resistance
   B. Voltage
   C. Current draw
   D. Polarity

12. A turbocharger fails frequently due to bearing failure. Technician A says oil contamination from turbo boost buildup in the engine could be the cause. Technician B says restricted oil passages to the turbocharger could be the problem. Who is correct?

    A. A only
    B. B only
    C. Both A and B
    D. Neither A nor B

13. Which of the following is the first noticeable symptom of a restricted air filter?

    A. Loss of power
    B. Decrease in fuel economy
    C. A no-start condition
    D. Excessive oil consumption.

14. Technician A says to always check the exhaust passages when replacing an EGR valve. Technician B says an EGR valve lowers carbon monoxide emissions. Who is correct?

    A. A only
    B. B only
    C. Both A and B
    D. Neither A nor B

15. Technician A says when completing monitor readiness, the vehicle must be driven under certain specific conditions in order for the monitors to run. Technician B says the emissions system that is being monitored must not have any related DTCs stored. Who is correct?

    A. A only
    B. B only
    C. Both A and B
    D. Neither A nor B

16. A 5-gas exhaust emissions analyzer may be used to help diagnose all of the following problems EXCEPT:

    A. Cylinder misfire
    B. Stuck-open injector
    C. Burnt valve
    D. Lazy oxygen sensor

17. When inspecting spark plugs, which one of the following would indicate a detonation problem?

    A. White insulator tip
    B. Wet carbon deposits on the insulator tip
    C. Black insulator tip
    D. Broken insulator tip

18. Which of the following is the LEAST LIKELY cause of poor fuel mileage on a vehicle with SFI?

    A. Restricted fuel pressure regulator
    B. Fuel pressure regulator partially stuck open
    C. Injector stuck open
    D. Burnt valve

19. Which of the follow is a function of an AIR system diverter valve?

    A. Lean out the mixture on deceleration.
    B. Richen the mixture on deceleration.
    C. Prevent a backfire on deceleration.
    D. Pull fresh air into the exhaust on deceleration.

20. Which of the following computer sensor signals is not used during open-loop mode?

    A. Manifold absolute-pressure sensor
    B. Throttle position sensor
    C. Engine-coolant temperature sensor
    D. Heated oxygen sensor

21. Which of the following is the LEAST LIKELY cause of low cylinder compression?

    A. Burnt valve
    B. Worn valve guides
    C. Worn piston rings
    D. Worn valve seats

22. A starter free-spinning test is being done on the bench with a fully charged battery. The current draw is higher than specification and the RPM is lower. Technician A says this could be caused by tight bushings. Technician B says typical current draw when free spinning a starter is 60 to 100 amps. Who is correct?

    A. A only
    B. B only
    C. Both A and B
    D. Neither A nor B

23. The secondary ignition system can be tested using which device?

    A. Test light
    B. Ohmmeter
    C. An ammeter
    D. Logic probe

24. Which of these steps should the technician take first when faulty fuel pump pressure is suspected?

    A. Check the fuel filter for restriction
    B. Check for kinked fuel lines
    C. Check fuel pressure and volume
    D. Check for fuel related diagnostic trouble codes

25. What should the fuel level be before an evaporative emissions monitor will run?

    A. 50 percent
    B. Over 25 percent
    C. At least 75 percent
    D. Between 15 percent and 85 percent

26. A vehicle has a misfire diagnostic trouble code. Technician A says the PCM uses the camshaft sensor signal to monitor engine misfire. Technician B says a misfire diagnostic code is a high-priority code and should be diagnosed first before other non-emissions-related codes. Who is correct?

    A. A only

    B. B only

    C. Both A and B

    D. Neither A nor B

27. Technician A says catalytic converter efficiency is not monitored on OBD-II compliant vehicles. Technician B says some vehicles may have no computer diagnostic capabilities for the secondary air-injection system. Who is correct?

    A. A only

    B. B only

    C. Both A and B

    D. Neither A nor B

28. A vehicle with double overhead valves is being diagnosed for backfiring through the exhaust manifold. Technician A says this engine uses a separate camshaft for the intake and exhaust valves. Technician B says a 4-cylinder double overhead camshaft (DOHC) engine has two camshafts. Who is correct?

    A. A only

    B. B only

    C. Both A and B

    D. Neither A nor B

29. A vehicle with a hard-starting complaint is being diagnosed. The battery shows an open circuit voltage of 11.7 volts. Technician A says the battery should be recharged then load-tested. Technician B says the battery cranking voltage should also be checked. Who is correct?

    A. A only

    B. B only

    C. Both A and B

    D. Neither A nor B

30. High secondary ignition-system circuit resistance can be caused by all of the following EXCEPT:

    A. Excessive spark plug gap

    B. Spark plug gap bridging

    C. Open circuit in the ignition coil

    D. Excessive distributor cap terminal to rotor gap

31. A vacuum assisted fuel pressure regulator is used on sequential fuel injection for which of the following reasons?

    A. To increase fuel delivery under high load conditions

    B. To provide a constant pressure drop across the injector due to a rapidly opening throttle

    C. To improve injector spray patterns

    D. To prevent fuel-pressure leak down when the engine is turned off

32. Technician A says some vehicles illuminate a loose gas cap light and store a code if a very large leak is detected. Technician B says the vent valve is a normally closed solenoid. Who is correct?

   A. A only
   B. B only
   C. Both A and B
   D. Neither A nor B

33. The cause of a malfunctioning indicator lamp (MIL) not illuminating during a bulb check when the ignition is turned on can be all of the following EXCEPT:

   A. A blown bulb
   B. An internal circuit problem in the PCM
   C. A open in the MIL circuit
   D. A diagnostic trouble code stored in the PCM

34. Which of the following is the LEAST LIKELY cause of blue exhaust smoke from the engine?

   A. Worn piston rings
   B. Worn valve guides
   C. Worn valve seats
   D. Worn valve seals

35. Technician A says that the ignition module controls the timing during start up on some ignition systems. Technician B says that on some ignition systems, the PCM has full control of timing at all times. Who is correct?

   A. A only
   B. B only
   C. Both A and B
   D. Neither A nor B

36. A vehicle with sequential fuel injection is running rough. A lab scope shows all injector waveforms to be identical except for one that has a considerably shorter voltage spike than the others. Which of the following is the most likely cause?

   A. Open connection at the injector
   B. Bad PCM
   C. Shorted injector winding
   D. Low charging system voltage

37. Technician A says if the exhaust gas recirculation (EGR) valve remains open at idle and low speed, then detonation can occur. Technician B says if the EGR valve does not open at cruising speeds, then the vehicle will run rough. Who is correct?

   A. A only
   B. B only
   C. Both A and B
   D. Neither A nor B

38. A soft code in the PCM memory is a code that:

    A. Indicates a top priority code
    B. Occurred in the past, but no longer exists
    C. Exists at the time the vehicle is being tested
    D. Can be found if a diagnostic flow chart is used

39. A vehicle with a slow cranking complaint has a voltmeter connected to the 12-volt battery. With the engine cranking, what is the lowest recommended voltmeter reading?

    A. 9.6 volts
    B. 11.0 volts
    C. 7.5 volts
    D. 12.0 volts

40. Technician A says the crankshaft sensor may be rotated to adjust the base ignition timing on some engines with electronic ignition. Technician B says on some systems, the crankshaft sensor interrupter ring is part of the crankshaft. Who is correct?

    A. A only
    B. B only
    C. Both A and B
    D. Neither A nor B

41. Technician A says some vehicles are more sensitive to aftermarket high-performance air filters than others. Technician B says to always check technical service bulletins when an aftermarket high-performance air filter has been installed and a drivability complaint is being diagnosed. Who is correct?

    A. A only
    B. B only
    C. Both A and B
    D. Neither A nor B

42. While testing a pulse-air-type air-injection system, Technician A says with the air cleaner's fresh-air hose removed and the engine idling, there should be steady audible pulses at the air cleaner inlet. Technician B says on some vehicles, air pulses should be felt escaping through the air cleaner inlet with the fresh air hose removed. Who is correct?

    A. A only
    B. B only
    C. Both A and B
    D. Neither A nor B

43. Technician A says freeze frame data is always set for a misfire code. Technician B says a misfire code is considered a high-priority code. Who is correct?

    A. A only
    B. B only
    C. Both A and B
    D. Neither A nor B

44. An oxygen sensor reads higher-than-normal voltage. Technician A says that the engine may have a vacuum leak. Technician B says that the exhaust manifold could be leaking. Who is correct?

    A. A only
    B. B only
    C. Both A and B
    D. Neither A nor B

45. A fully charged 12-volt battery should measure what voltage?

    A. 12.0 volts
    B. 12.3 volts
    C. 12.5 volts
    D. 12.6 volts

46. An open spark plug wire was found on an engine that was missing on acceleration. Technician A says that the distributor cap and rotor should be carefully inspected for carbon tracks. Technician B says that the ignition coil should be replaced because the open wire could have caused the coil to become tracked internally. Who is correct?

    A. A only
    B. B only
    C. Both A and B
    D. Neither A nor B

47. Using which of the following can an injector pulse can be tested at the injector connector?

    A. Neon spark tester
    B. Vacuum hose and a test light
    C. DVOM
    D. NOID light

48. An evaporative emissions system that does not purge correctly can cause all of the following EXCEPT:

    A. Loss of fuel mileage
    B. Increased crankcase blowby
    C. Rich air/fuel ration
    D. An increase in tail pipe emissions

49. A short-term fuel trim of 0 percent and a long-term fuel trim of +19 percent means what?

    A. The engine has a history of running rich.
    B. The engine is running lean at the present time.
    C. The engine is running rich at the present time.
    D. The engine has a history of running lean.

50. A typical oxygen sensor output signal can be measured with the DVOM set on what unit of measure?

    A. AC volts
    B. DC volts
    C. Ohms
    D. Frequency

INTRODUCTION

Included in this section are the answer keys for each preparation exam, followed by individual, detailed answer explanations and a reference identifying the designated task area being assessed by each specific question. This additional reference information may prove useful if you need to refer back to the task list located in Section 4 of this book for additional support.

PREPARATION EXAM 1—ANSWER KEY

1.	B	21.	A	41.	D
2.	C	22.	C	42.	B
3.	D	23.	A	43.	C
4.	B	24.	D	44.	C
5.	C	25.	C	45.	D
6.	C	26.	A	46.	C
7.	A	27.	A	47.	B
8.	C	28.	B	48.	A
9.	A	29.	D	49.	C
10.	C	30.	C	50.	B
11.	C	31.	C		
12.	D	32.	B		
13.	C	33.	A		
14.	C	34.	C		
15.	B	35.	C		
16.	A	36.	A		
17.	C	37.	D		
18.	C	38.	D		
19.	C	39.	C		
20.	B	40.	B		

PREPARATION EXAM 1—EXPLANATIONS

1. Technician A says valve adjustment should always be performed on a cold engine. Technician B says the piston should be placed at top dead center (TDC) of the compression stroke. Who is correct?

    A. A only
    B. B only
    C. Both A and B
    D. Neither A nor B

    TASK A.11

    Answer A is incorrect. While many manufacturers require the engine to be cold when the valves are adjusted, some manufacturers specify the valves be adjusted on a hot engine.

    Answer B is correct. Only Technician B is correct. The correct piston position for valve adjustment is at TDC on the compression stroke. This ensures the valves are closed completely.

    Answer C is incorrect. Only Technician B is correct.

    Answer D is incorrect. Technician B is correct.

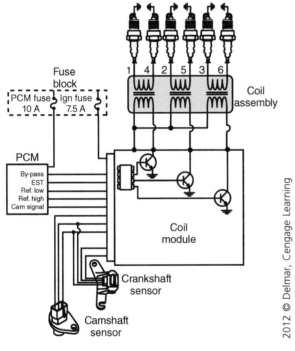

2. Refer to the illustration. Technician A says a shorted coil could affect two cylinders. Technician B says the schematic shows a waste spark ignition system. Who is correct?

    A. A only
    B. B only
    C. Both A and B
    D. Neither A nor B

    TASK B.5

    Answer A is incorrect. Technician B is also correct.

    Answer B is incorrect. Technician A is also correct.

    Answer C is correct. Both Technicians are correct. The illustration shows a waste spark system. In a waste spark ignition system, the secondary voltage jumps the gap of one spark plug, travels through the block, and then jumps the gap of the other spark plug.

    Answer D is incorrect. Both Technicians are correct.

TASK C.6

3. Technician A says fuel pressure readings that are above specifications could be caused by a stuck-open fuel pressure regulator. Technician B says if fuel is present at the vacuum hose port of a fuel pressure regulator, then the regulator is stuck open. Who is correct?

   A. A only
   B. B only
   C. Both A and B
   D. Neither A nor B

   Answer A is incorrect. A stuck-open fuel pressure regulator would cause fuel pressure to be lower than specifications, not higher.

   Answer B is incorrect. Fuel present at the vacuum port of a fuel pressure regulator is a sign of a ruptured diaphragm, and the regulator would need replacing.

   Answer C is incorrect. Neither Technician is correct.

   Answer D is correct. Neither Technician is correct. Fuel pressure higher than specifications can be caused by a stuck-closed fuel pressure regulator or a restriction in the return line on a return-type fuel system.

TASK D.1.2

4. Technician A says if the positive crankcase ventilation (PCV) valve rattled when shaken, then the PCV system is OK. Technician B says the PCV system vents excess pressure formed in the crank case from piston blow-by. Who is correct?

   A. A only
   B. B only
   C. Both A and B
   D. Neither A nor B

   Answer A is incorrect. While rattling the PCV valve is a way of testing the valve, a rattling valve does not mean the PCV system is working. It simply means the plunger in the valve is free to move.

   Answer B is correct. Only Technician B is correct. The PCV system is used to vent excess pressure in the crankcase through the engine. The excess pressure develops when compression pressures escape past the piston rings on the compression stroke. An inoperative PCV system can lead to oil leaks, premature gasket, and seal failure.

   Answer C is incorrect. Only Technician B is correct.

   Answer D is incorrect. Technician B is correct.

TASK E.5

5. Technician A says a greenish corrosion on terminals results in high resistance in the circuit. Technician B says loose retaining lock tabs on a terminal can cause high resistance in connector terminals. Who is correct?

   A. A only
   B. B only
   C. Both A and B
   D. Neither A nor B

   Answer A is incorrect. Technician B is also correct.

   Answer B is incorrect. Technician A is also correct.

   Answer C is correct. Both Technicians are correct. A greenish corrosion known as *green death* results from moisture at a terminal. If left unattended long enough, the terminal will be completely corroded away. This causes a high resistance in the circuit. Another common cause of high resistance in a wiring circuit is a broken terminal lock tab, which could cause the terminal to push out of the connector and become disconnected.

   Answer D is incorrect. Both Technicians are correct.

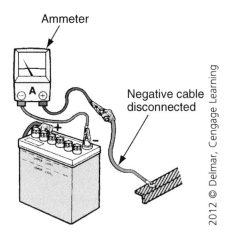

Ammeter

Negative cable disconnected

2012 © Delmar, Cengage Learning

6. Technician A says a key-off current draw test is being performed in the above illustration. Technician B says the allowable reading for this test is less than .05 amps. Who is correct?

TASK A.17

A. A only

B. B only

C. Both A and B

D. Neither A nor B

Answer A is incorrect. Technician B is also correct.

Answer B is incorrect. Technician A is also correct.

Answer C is correct. Both Technicians are correct. Key-off current draw, called a *parasitic load*, is being tested and should be less than .05 amps.

Answer D is incorrect. Both Technicians are correct.

7. Technician A says a single-cylinder misfire diagnostic trouble code (DTC) can be caused by a defective coil on some vehicles. Technician B says a low fuel pump pressure can cause a single-cylinder misfire diagnostic trouble code. Who is correct?

TASK B.2

A. A only

B. B only

C. Both A and B

D. Neither A nor B

Answer A is correct. Only Technician A is correct. Some vehicles use coil-on-plug (COP) ignition systems. On these ignition systems, each cylinder has its own ignition coil.

Answer B is incorrect. Low fuel pressure would cause multiple cylinder misfires, not a single cylinder misfire, because low fuel pressure would affect all the cylinders.

Answer C is incorrect. Only Technician A is correct.

Answer D is incorrect. Technician A is correct.

TASK C.4

8. Technician A says a vehicle with a restricted fuel filter can still have fuel pressure within the specifications. Technician B says if a restricted fuel filter is suspected, then a fuel pump volume test can be performed. Who is correct?

   A.   A only
   B.   B only
   C.   Both A and B
   D.   Neither A nor B

   Answer A is incorrect. Technician B is also correct.

   Answer B is incorrect. Technician A is also correct.

   Answer C is correct. Both Technicians are correct. While a restricted fuel filter will affect the amount (volume) of fuel that can be pumped, it will not affect pressure.

   Answer D is incorrect. Both Technicians are correct.

TASK D.2.1

9. All of the following can cause the engine to spark knock EXCEPT:

   A.   A stuck-open exhaust recirculation valve (EGR)
   B.   A stuck-closed EGR
   C.   A broken vacuum hose going to the EGR
   D.   A defective cooling system thermostat

   Answer A is correct. The EGR system is used to control combustion chamber temperature, by redirecting inert exhaust gas by into the cylinder. An EGR valve stuck open would cause a rough idle or maybe a stall out at idle, but it would not cause spark knock.

   Answer B is incorrect. If the EGR valve were to stick closed, then the combustion chamber could reach temperatures high enough to cause spark knock to occur.

   Answer C is incorrect. A broken EGR hose would prevent the ERG valve from opening, causing excessive combustion chamber temperature, which could cause spark knock.

   Answer D is incorrect. A defective cooling system thermostat could cause the engine to operate at a higher than normal temperature, this could cause spark knock in the engine.

TASK E.7

10. Technician A says when replacing the powertrain control module (PCM), the new PCM may have to be reprogrammed in order for it to operate. Technician B says some drivability problems can be fixed by reprogramming the PCM. Who is correct?

    A.   A only
    B.   B only
    C.   Both A and B
    D.   Neither A nor B

    Answer A is incorrect. Technician B is also correct.

    Answer B is incorrect. Technician A is also correct.

    Answer C is correct. Both Technicians are correct. On many vehicles, the PCM must be reprogrammed when it is replaced. Sometimes a reprogramming (re-flashing) is necessary to correct drivability complaints.

    Answer D is incorrect. Both Technicians are correct.

11. Technician A says one of the first steps in diagnosing a drivability complaint is to verify the driver's complaint. Technician B says one of the first steps in diagnosing a drivability complaint is to perform a thorough visual inspection. Who is correct?

TASK A.1

    A. A only

    B. B only

    C. Both A and B

    D. Neither A nor B

Answer A is incorrect. Technician B is also correct.

Answer B is incorrect. Technician A is also correct.

Answer C is correct. Both Technicians are correct. When diagnosing drivability complaints, the first step is to verify the customer's complaint. Sometimes problems can be fixed by simply performing a good visual inspection. Problems like cracked vacuum hoses, wiring problems, and leaks often can be found with a visual check.

Answer D is incorrect. Both Technicians are correct.

12. Technician A says the normal secondary-circuit resistance of an ignition coil is low (5 ohms or less). Technician B says the spark plug wire is a secondary component that should have a resistance of 20,000 ohms or more per foot. Who is correct?

TASK B.4

    A. A only

    B. B only

    C. Both A and B

    D. Neither A nor B

Answer A is incorrect. The secondary circuit in the ignition coil will have a high resistance, around 9,000 to 30,000 ohms.

Answer B is incorrect. Good spark plug wires will measure less than 10,000 ohms per foot.

Answer C is incorrect. Neither Technician is correct.

Answer D is correct. Neither Technician is correct. When testing the resistance of an ignition coil, the primary circuit should be around 1 to 3 ohms; however, some coils have an even lower specification. The secondary circuit resistance should be around 9,000 to 30,000 ohms depending on the manufacturer. Spark plug wires should be visually inspected and their resistance checked. Resistance is usually less than 10,000 ohms per foot.

13. Technician A says a vacuum leak can occur under the intake manifold and cause oil consumption. Technician B says propane is a good method for locating vacuum leaks. Who is correct?

TASK C.10

    A. A only

    B. B only

    C. Both A and B

    D. Neither A nor B

Answer A is incorrect. Technician B is also correct.

Answer B is incorrect. Technician A is also correct.

Answer C is correct. Both Technicians are correct. A vacuum leak under the intake manifold can be hard to diagnose. A good indication of a vacuum leak under the intake manifold is the complaint of excessive oil consumption coupled with a drivability complaint. A can of propane connected by an adapter to a long hose can help to access hard-to-reach places. When checking for vacuum leaks, note that the engine RPM will rise or fall depending on the air/fuel mixture at the time the leak is found.

Answer D is incorrect. Both Technicians are correct.

TASK D.3.3

14. Technician A says a secondary air-injection system directs output from the air pump to the exhaust manifold during engine warm-up and switches air to the catalytic converter during closed-loop operation. Technician B says air injection will have little or no effect on tailpipe carbon dioxide readings. Who is correct?

   A. A only
   B. B only
   C. Both A and B
   D. Neither A nor B

   Answer A is incorrect. Technician B is also correct.

   Answer B is incorrect. Technician A is also correct.

   Answer C is correct. Both Technicians are correct. During warm-up, the secondary air is directed upstream to the exhaust manifold. After the vehicle enters closed-loop mode, the secondary air is directed downstream to the catalytic converter. A carbon dioxide reading indicates engine efficiency and is not affected by the secondary air pump.

   Answer D is incorrect. Both Technicians are correct.

TASK E.9

15. Technician A says the best way to clear diagnostic trouble codes is to remove the battery negative terminal for five seconds. Technician B says the monitor for the repaired system should be performed before the vehicle is returned to the customer. Who is correct?

   A. A only
   B. B only
   C. Both A and B
   D. Neither A nor B

   Answer A is incorrect. While some manufacturers did disconnect the battery to clear trouble codes many years ago, this is not the best method today. The most preferred method is to use the scan tool, then the PCM fuse, and last the battery, if no other way is allowed.

   Answer B is correct. Only Technician B is correct. With On-Board Diagnostics II (OBD II), the repair can be confirmed by driving the vehicle through the necessary drive cycles, running the system monitor on the system that was repaired.

   Answer C is incorrect. Only Technician B is correct.

   Answer D is incorrect. Technician B is correct.

TASK A.8

16. All of the following are true of the cylinder leakage test EXCEPT:

   A. Air loss and bubbles in the radiator indicate a bad intake valve guide.
   B. Air loss from the oil filler cap indicates worn piston rings.
   C. A gauge reading of 0 percent indicates no cylinder leakage.
   D. Air loss from the exhaust indicates a valve problem.

   Answer A is correct. When performing a cylinder leakage test, air is injected into the cylinder with the piston at TDC on the compression stroke. Air loss or bubbles in the radiator indicate a bad head gasket or cracked head or block.

   Answer B is incorrect. Air loss from the oil cap would indicate air leaking past the piston rings.

   Answer C is incorrect. A reading of 0 percent indicates no cylinder leakage. Most engines in good shape have a leakage of 10 percent or less.

   Answer D is incorrect. Air leaking from the exhaust indicates an exhaust valve problem.

17. Technician A says a faulty cam shaft position sensor can cause a no-start condition. Technician B says some vehicles will start without an operating camshaft position sensor (CMP). Who is correct?

TASK B.1

A. A only

B. B only

C. Both A and B

D. Neither A nor B

Answer A is incorrect. Technician B is also correct.

Answer B is incorrect. Technician A is also correct.

Answer C is correct. Both Technicians are correct. The CMP is used in conjunction with the crankshaft position sensor (CKP) to detect which cylinder is firing for spark and as a reference to time the sequential fuel injection. Most vehicles will not start without the CMP signal; however, some vehicles will.

Answer D is incorrect. Both Technicians are correct.

18. Technician A says the exhaust can be checked for restrictions using a vacuum gauge. Technician B says the back pressure in the exhaust should not exceed 1.5 psi at idle. Who is correct?

TASK C.13

A. A only

B. B only

C. Both A and B

D. Neither A nor B

Answer A is incorrect. Technician B is also correct.

Answer B is incorrect. Technician A is also correct.

Answer C is correct. Both Technicians are correct. A vacuum gauge is used to test for exhaust restriction. A vehicle with restricted exhaust will have a low vacuum reading at 2,000 rpm and possibly slowly drop even lower as the RPM is held. The best method for checking the exhaust for restriction is to check exhaust back pressure, which should not exceed 1.5 psi at idle and 2.5 psi at 2,000 rpm.

Answer D is incorrect. Both Technicians are correct.

19. Technician A says a leaking gas cap gasket can cause an evaporative emissions failure. Technician B says some evaporative emissions canisters have a replaceable filter. Who is correct?

TASK D.4.3

A. A only

B. B only

C. Both A and B

D. Neither A nor B

Answer A is incorrect. Technician B is also correct.

Answer B is incorrect. Technician A is also correct.

Answer C is correct. Both Technicians are correct. The purpose of the evaporative emissions system is to prevent the release of gasoline vapors into the atmosphere. A leaking gas cap will cause the malfunctioning indicator lamp to illuminate on a vehicle with enhanced EVAP system. On older vehicles the vapor canisters use a fiberglass-type canister filter that is replaced during maintenance service intervals.

Answer D is incorrect. Both Technicians are correct.

TASK E.1

20. Technician A says a diagnostic trouble code (DTC) tells you which component is malfunctioning. Technician B says if a fault exists that affects emissions, then the malfunction indicator lamp (MIL) will be illuminated. Who is correct?

   A. A only
   B. B only
   C. Both A and B
   D. Neither A nor B

   Answer A is incorrect. A DTC is set when a system or circuit is out of its parameters. Once a DTC is set the technician uses the DTC to choose which path to take toward diagnosis. A diagnostic flow chart is used to test certain components related to the system or circuit. No component should be replaced solely based on a DTC.

   Answer B is correct. Only Technician B is correct. The MIL will illuminate any time a failure of a circuit or system occurs that causes the vehicle to emit emissions 1.5 times the federal test procedure allowance.

   Answer C is incorrect. Only Technician B is correct.

   Answer D is incorrect. Technician B is correct.

TASK A.4

21. Technician A says blue smoke in the exhaust indicates oil being burned in the combustion chamber. Technician B says white smoke in the exhaust indicates a rich air/fuel ratio. Who is correct?

   A. A only
   B. B only
   C. Both A and B
   D. Neither A nor B

   Answer A is correct. Only Technician A is correct. A vehicle that is burning oil will have a blue to gray color to the exhaust.

   Answer B is incorrect. White smoke is an indication of coolant in the exhaust; some white exhaust is normal, however, depending on the climate and temperature. A vehicle that is running too rich will have black exhaust.

   Answer C is incorrect. Only Technician A is correct.

   Answer D is incorrect. Technician A is correct.

TASK B.3

22. All of the following are components of the primary circuit EXCEPT:

   A. Ignition coil
   B. Ignition switch
   C. Spark plug wire
   D. Battery

   Answer A is incorrect. The ignition coil has a primary circuit and a secondary circuit.

   Answer B is incorrect. The ignition switch is one of the primary ignition components that opens and closes the ignition system.

   Answer C is correct. The spark plug wire is a secondary ignition component; secondary ignition components handle high voltage (20,000–50,000 volts) and primary handle low voltage (12 volts).

   Answer D is incorrect. The battery is the first component of the primary ignition system, supplying the primary voltage.

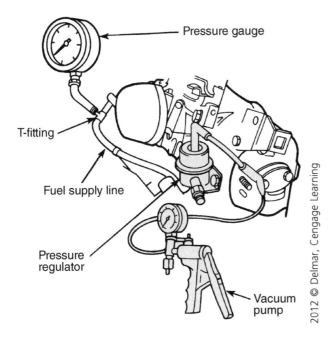

Pressure gauge

T-fitting

Fuel supply line

Pressure regulator

Vacuum pump

2012 © Delmar, Cengage Learning

23. Refer to the above illustration. Technician A says fuel pressure regulator operation is being tested. Technician B says as vacuum is applied to the regulator, the fuel pressure should increase. Who is correct?

TASK C.11

   A. A only
   B. B only
   C. Both A and B
   D. Neither A nor B

Answer A is correct. Only Technician A is correct. The fuel pressure regulator is being tested.

Answer B is incorrect. As the vacuum increases, the fuel pressure drops. The fuel pressure regulator is used to control the fuel pressure. It is located on the return side of the fuel system on a return-type fuel system and at the fuel tank on most returnless systems. On the type illustrated, as the throttle opens, the vacuum drops; this drop in vacuum causes the fuel pressure to increase a few pounds to help prevent a fuel pressure drop on acceleration.

Answer C is incorrect. Only Technician A is correct.

Answer D is incorrect. Technician A is correct.

24. Technician A says the PCV valve flow is high at idle. Technician B says PCV valve flow is high at cruising speed on a vehicle with an 8-cylinder engine. Who is correct?

TASK D.1.1

   A. A only
   B. B only
   C. Both A and B
   D. Neither A nor B

Answer A is incorrect The PCV valve flow is low at idle speeds.

Answer B is incorrect. When traveling at cruising speeds, the PCV valve flow is low.

Answer C is incorrect. Neither Technician is correct.

Answer D is correct. Neither Technician is correct. When manifold vacuum is high (at idle, cruising and light-load operation), the PCV valve restricts the airflow to maintain a balanced air-fuel ratio. Under high speeds or heavy loads, the PCV valve opens to allow maximum flow.

TASK E.2

25. A vehicle with a MIL illuminated is being diagnosed. The DTC stored in the PCM is PO300 (Random/Multiple-Cylinder Misfire Detected). Technician A says this is a one-trip failure, because catalyst damage can occur. Technician B says the MIL will flash on this type of code. Who is correct?

    A.   A only

    B.   B only

    C.   Both A and B

    D.   Neither A nor B

Answer A is incorrect. Technician B is also correct.

Answer B is incorrect. Technician A is also correct.

Answer C is correct. Both Technicians are correct. PO3XX codes are ignition system or misfire codes. A cylinder miss can damage the catalytic converter. If a cylinder miss is detected, then the MIL will begin flashing to alert the driver.

Answer D is incorrect. Both Technicians are correct.

TASK A.2

26. Technician A says that cranking the engine with the throttle fully depressed will force a lean mixture to clear a flooded engine on some vehicles. Technician B says that high system voltage will increase fuel injector on time (pulse width). Who is correct?

    A.   A only

    B.   B only

    C.   Both A and B

    D.   Neither A nor B

Answer A is correct. Only Technician A is correct. Depressing the throttle fully during cranking will make the computer enter clear flood mode and lean out the air/fuel mixture on some vehicles.

Answer B is incorrect. Low, not high, system voltage will enable battery voltage correction programming in the computer, which will increase injector pulse-width commands due to slower injector opening from low system voltage.

Answer C is incorrect. Only Technician A is correct.

Answer D is incorrect. Technician A is correct.

TASK B.5

27. Which of the following is the most likely cause of ignition coil failure?

    A.   Prolonged open circuit in the secondary

    B.   Open ignition module primary circuit

    C.   Short-to-ground in the secondary

    D.   Open in the trigger device circuit

Answer A is correct. An open in the secondary circuit can cause internal tracking in the ignition coil, causing premature coil failure.

Answer B is incorrect. An open in the primary circuit of the ignition module will cause a no-start, but will not cause ignition coil failure.

Answer C is incorrect. A short-to-ground in the secondary can cause a miss or a no-start, but will not cause ignition coil failure.

Answer D is incorrect. An open in the trigger device circuit will cause a no-start, but will not cause ignition coil failure.

28. Technician A says performing a fuel pressure test confirms proper operation of the fuel injector. Technician B says it is possible to have an electrical problem with an injector, even though the fuel pressure drop is within specifications. Who is correct?

TASK C.8

   A. A only

   B. B only

   C. Both A and B

   D. Neither A nor B

   Answer A is incorrect. A fuel pressure test confirms the fuel pump and regulator are controlling the pressure correctly. It does not confirm adequate volume or injector operation.

   Answer B is correct. Only Technician B is correct. The fuel pressure drop test is used to check fuel flow through the injector, not the electrical circuit. The injector must be checked for resistance to check the electrical function.

   Answer C is incorrect. Only Technician B is correct.

   Answer D is incorrect. Technician B is correct.

29. Technician A says the EGR system is used to raise combustion chamber temperature. Technician B says that EGR systems that use an EGR valve position sensor should read about .7 volts with the EGR valve at full open. Who is correct?

TASK D.2.2

   A. A only

   B. B only

   C. Both A and B

   D. Neither A nor B

   Answer A is incorrect. The EGR system is used to lower combustion chamber temperature so that NOx is reduced.

   Answer B is incorrect. An EGR valve position sensor will read close to 5 volts at full open.

   Answer C is incorrect. Neither Technician is correct.

   Answer D is correct. Neither Technician is correct. The EGR system is used to lower combustion chamber temperature in order to limit NOx formation in the engine. Some manufacturers use an EGR position sensor for feedback of EGR pintle position.

30. A scan test of the computer system on a late-model fuel-injected engine reveals a bank #1 long-term fuel trim value of –19, and a bank #2 long-term fuel trim value of –18 with the engine idling. Technician A says these readings could be caused by a defective fuel pressure regulator. Technician B says a restricted fuel return line could cause these readings. Who is correct?

TASK E.4

   A. A only

   B. B only

   C. Both A and B

   D. Neither A nor B

   Answer A is incorrect. Technician B is also correct.

   Answer B is incorrect. Technician A is also correct.

   Answer C is correct. Both Technicians are correct. The fuel trim readings show that the PCM is subtracting fuel, so it is leaning down the air/fuel mixture. A defective fuel pressure regulator can cause these types of fuel trim readings if it sticks and causes high fuel pressure. A restricted fuel return line will cause high fuel pressures, resulting in a negative fuel trim.

   Answer D is incorrect. Both Technicians are correct.

TASK A.3

31. Technician A says a stethoscope can be used to pinpoint engine noises. Technician B says you may use a long screwdriver for noise diagnosis if a stethoscope is not available. Who is correct?

   A. A only
   B. B only
   C. Both A and B
   D. Neither A nor B

   Answer A is incorrect. Technician B is also correct.

   Answer B is incorrect. Technician A is also correct.

   Answer C is correct. Both Technicians are correct. A stethoscope is a good tool to use to help diagnose an engine noise; if a stethoscope is not available, then a long screwdriver can be used to listen to noises.

   Answer D is incorrect. Both Technicians are correct.

TASK A.13

32. Engine oil does all of the following tasks EXCEPT:

   A. Controls the variable valve timing system
   B. Helps cool the charging system
   C. Helps seal the piston rings
   D. Lubricates cam phaser

   Answer A is incorrect. Failure to change it at regular maintenance intervals and at the correct level and viscosity can cause problems with the oil-controlled variable valve timing system.

   Answer B is correct. Engine oil helps maintain many systems, but is not involved with the electrical charging system.

   Answer C is incorrect. Along with controlling the variable valve timing system on some vehicles, engine oil helps cool the engine, reduces friction, cleans and holds dirt, and helps seal the piston rings.

   Answer D is incorrect. Oil that becomes dirty and sludgy can cause the cam phaser and oil control valves to stick, preventing the intake and exhaust valves from operating at different points in the combustion cycle to improve engine performance.

TASK C.1

33. Technician A says that turning off fuel injectors at high RPM is the purpose of the rev limiter—to protect the engine from damage or limit vehicle speed. Technician B says that turning off fuel injectors while the engine is running must be done at speeds above 45 mph. Who is correct?

   A. A only
   B. B only
   C. Both A and B
   D. Neither A nor B

   Answer A is correct. Only Technician A is correct. The rev limiter is used to protect the engine from over revving. Rev limiters are set at different RPMs depending on manufacturer.

   Answer B is incorrect. Most fuel-injection systems turn off fuel injectors under certain conditions to reduce fuel consumption and emissions and to protect the catalytic converter from over-rich mixtures at any speed.

   Answer C is incorrect. Only Technician A is correct.

   Answer D is incorrect. Technician A is correct.

34. Technician A says to use an anti-backfire valve to prevent backfiring during deceleration on pump-driven air-injection systems. Technician B says to prevent exhaust gases from back-flowing into the air-injection control valves or pump, check valves can be used in the exhaust manifold and converter feed pipes. Who is correct?

TASK D.3.1

    A. A only

    B. B only

    C. Both A and B

    D. Neither A nor B

Answer A is incorrect. Technician B is also correct.

Answer B is incorrect. Technician A is also correct.

Answer C is correct. Both Technicians are correct. The anti-backfire valve, sometimes called the *diverter valve*, helps prevent a backfire in the exhaust on deceleration. To protect the secondary injection system from hot exhaust, check valves are used. These valves allow air in from the air pump but no exhaust out to the secondary air-injection components.

Answer D is incorrect. Both Technicians are correct.

35. A MAF (maximum air flow) load-calculating-type port fuel-injected engine runs fine at idle, but hesitates under acceleration. No DTCs are stored. Technician A says to check for a restricted MAF inlet screen. Technician B says a bad spot in the throttle position sensor signal could be the cause. Who is correct?

TASK E.3

    A. A only

    B. B only

    C. Both A and B

    D. Neither A nor B

Answer A is incorrect. Technician B is also correct.

Answer B is incorrect. Technician A is also correct.

Answer C is correct. Both Technicians are correct. A restricted MAF inlet screen will cause an improper response from the sensor and a hesitation under acceleration. A TPS signal that drops out will cause the computer to reduce fuel flow thinking the throttle was closing, causing a hesitation under acceleration.

Answer D is incorrect. Both Technicians are correct.

36. Which of the following is not a cause of a noisy valve train?

TASK A.3

    A. High oil pressure

    B. Collapsed lifter

    C. Incorrect valve adjustment

    D. Bent pushrod

Answer A is correct. High oil pressure would not cause noise; however, low oil pressure can cause valve train noise.

Answer B is incorrect. A collapsed lifter would cause a noise similar to the sound of an ink pen clicking.

Answer C is incorrect. Incorrect valve adjustment will cause noise if the valves are adjusted too loosely.

Answer D is incorrect. A bent pushrod will cause valve train noise because of the extra clearance the bent pushrod causes.

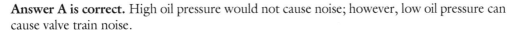

TASK B.1

37. A cylinder misfire will cause all the following EXCEPT:

A. High hydrocarbon (HC) emissions
B. High oxygen level in the exhaust
C. Damage to the catalytic converter
D. High carbon dioxide (CO_2) readings

Answer A is incorrect. HCs are nothing more than raw gasoline. A cylinder miss would allow the fuel to pass through the engine without being burned.

Answer B is incorrect. If a cylinder is missing, then the fuel is not being burned correctly. If the fuel is not burning, then it is not using the oxygen (O_2) in the missing cylinder, so the O_2 level would be high.

Answer C is incorrect. A cylinder missing will cause the catalytic converter to overheat very quickly causing converter damage.

Answer D is correct. CO_2 readings are engine efficiency indicators. A missing engine would have a low CO_2 reading.

TASK C.2

38. A vehicle is being diagnosed for a PO134 DTC (Oxygen Sensor Circuit, No Activity Detected) (bank #1, sensor #1). Technician A says the problem could be high fuel system pressure. Technician B says the problem is more likely confined to the downstream O_2 sensor on the bank containing cylinder #1. Who is correct?

A. A only
B. B only
C. Both A and B
D. Neither A nor B

Answer A is incorrect. Low fuel pressure would affect all sensor readings, not just one.

Answer B is incorrect. Bank #1, sensor #1 is the upstream sensor on the bank containing cylinder #1.

Answer C is incorrect. Neither Technician is correct.

Answer D is correct. Neither Technician is correct. The oxygen sensors typically are labeled as HO2S 1/1 (bank #1, sensor #1), HO2S 1/2 (bank #1, sensor #2), HO2S 2/1 (bank #2, sensor #1). Bank #1 is located on the same side as cylinder #1. Sensor #1 is the upstream and sensor #2 is the downstream.

TASK D.2.1

39. Technician A says that an EGR valve stuck closed will cause detonation. Technician B says that an EGR valve stuck closed will cause high NOx (oxides of nitrogen) emissions. Who is correct?

A. A only
B. B only
C. Both A and B
D. Neither A nor B

Answer A is incorrect. Technician B is also correct.

Answer B is incorrect. Technician A is also correct.

Answer C is correct. Both Technicians are correct. The EGR valve is used to lower the formation of NOx. It does this by redirecting a controlled amount of exhaust gas to the intake manifold. The exhaust gas is inert, meaning it can't be re-burned, so the inert gas takes up space in the combustion chamber that would otherwise have fuel and air in it. This lowers the combustion chamber temperature, reducing the formation of NOx and the possibility of detonation.

Answer D is incorrect. Both Technicians are correct.

40. Technician A says that the resistance in the power side of the power distribution circuit cannot be over 200 ohms of resistance. Technician B says the power side of the power distribution circuit can be voltage-drop tested to check the circuit for resistance. Who is correct?

TASK E.6

   A. A only
   B. B only
   C. Both A and B
   D. Neither A nor B

   Answer A is incorrect. Any measurable resistance is not allowable in the power side of the power distribution circuit.

   Answer B is correct. Only Technician B is correct. Performing a voltage drop test is the best way to check for excessive resistance in an electrical circuit. Anything over 0.5 volts is considered high.

   Answer C is incorrect. Only Technician B is correct.

   Answer D is incorrect. Technician B is correct.

41. If the vacuum drops slowly to a low reading when a vacuum gauge is connected to the intake manifold and the engine is accelerated and held at a steady speed, then which of the following is the most likely cause?

TASK A.5

   A. A rich fuel mixture
   B. Over-advanced ignition timing
   C. Sticking valves
   D. A restricted exhaust

   Answer A is incorrect. A rich mixture would cause black smoke from the exhaust not the vacuum to slowly drop.

   Answer B is incorrect. Over-advanced timing would cause a spark knock condition and would not cause the vacuum to slowly drop.

   Answer C is incorrect. Sticking valves would cause a pulsating needle on the vacuum gauge.

   Answer D is correct. A restricted exhaust can be diagnosed using a vacuum gauge. With the gauge connected to the intake manifold, increase the RPM to 2,000 rpm and observe the gauge. If the vacuum slowly drops as the RPM is held at 2,000 rpm, then the exhaust could be restricted.

42. A vehicle with DTC code PO304 (Cylinder #4 Misfire) is being diagnosed. Technician A says high fuel pump pressure could be the cause. Technician B says an open spark plug wire could be the cause. Who is correct?

TASK B.2

   A. A only
   B. B only
   C. Both A and B
   D. Neither A nor B

   Answer A is incorrect. High fuel pressure would affect all the cylinders, not just one.

   Answer B is correct. Only Technician B is correct. An open spark plug wire, fouled spark plug, or shorted spark plug would all cause a misfire code.

   Answer C is incorrect. Only Technician B is correct.

   Answer D is incorrect. Technician B is correct.

TASK C.3

43. Which of the following is the LEAST LIKELY cause of a fuel tank leak?

 A. Rust
 B. Road damage
 C. Retaining straps
 D. Leaking tank seams

Answer A is incorrect. Rust and corrosion are often causes of a fuel tank leak, especially on older vehicles.

Answer B is incorrect. Road damage to fuel tanks is a common cause of fuel leaks.

Answer C is correct. Fuel tank retaining straps are very seldom the cause of fuel tank leaks.

Answer D is incorrect. Some leaks may occur along the seams of the fuel tank.

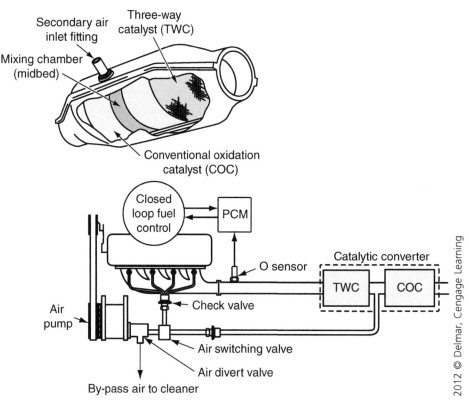

TASK D.3.3

44. Refer to the above illustration. Technician A says the air pump should pump the air into the exhaust manifold when in open loop. Technician B says the air should be pumped into the catalytic converter once the vehicle enters closed loop. Who is correct?

 A. A only
 B. B only
 C. Both A and B
 D. Neither A nor B

Answer A is incorrect. Technician B is also correct.

Answer B is incorrect. Technician A is also correct.

Answer C is correct. Both Technicians are correct. The secondary injection system is used to inject additional air into the catalytic converter so that the converters can operate. When the vehicle is in closed loop, the air is directed to the catalytic converter. When the vehicle is cold and in open loop, the air is sent to the exhaust manifold.

Answer D is incorrect. Both Technicians are correct.

45. A vehicle has a DTC for an open vehicle speed sensor circuit. Which of the following systems would most likely be affected?

    A. Air conditioning (A/C) system

    B. Manual transmission shift points

    C. Traction control system (TCS)

    D. Cruise control system

TASK E.8

Answer A is incorrect. The vehicle speed sensor would not affect the A/C system.

Answer B is incorrect. The vehicle speed sensor would not affect the manual transmission system.

Answer C is incorrect. The vehicle speed sensor would not affect the TCS. The TCS uses wheel-speed sensors for traction management.

Answer D is correct. The cruise control system uses the vehicle-speed sensor (VSS) for MPH input. Usually if a vehicle has a VSS DTC, then the cruise control also does not operate.

46. Technician A says the electrolyte level is important in a non-serviceable battery. Technician B says on some batteries, the electrolyte level can be checked in a sealed battery by looking through the translucent battery case. Who is correct?

    A. A only

    B. B only

    C. Both A and B

    D. Neither A nor B

TASK A.17

Answer A is incorrect. Technician B is also correct.

Answer B is incorrect. Technician A is also correct.

Answer C is correct. Both Technicians are correct. The level of the electrolytes is critical in any battery regardless of its serviceability. On some batteries, the case is translucent (clear) with the full line clearly marked, giving the technician the ability to quickly check the electrolyte level.

Answer D is incorrect. Both Technicians are correct.

47. What term describes the condition when the air/fuel mixture is ignited before the spark plug is fired?

    A. Dieseling

    B. Pre-ignition

    C. Lean ramping

    D. Early timing

TASK B.1

Answer A is incorrect. Dieseling occurs when the engine continues to run after the key is turned off. Dieseling is caused by the fuel being ignited by the heat of combustion.

Answer B is correct. Pre-ignition occurs when a hot spot in the combustion chamber causes the air/fuel mixture to ignite before the spark is delivered to the spark plug.

Answer C is incorrect. *Lean ramping* occurs when a vehicle leans out on deceleration to the point of causing the vehicle to surge.

Answer D is incorrect. Early timing, also called *advanced timing*, means the spark is delivered to the spark plug too soon.

TASK C.3

48. Technician A says that special fuel injection hose must be used when replacing the fuel lines on a fuel-injected vehicle. Technician B says the fuel supply hose is usually smaller than the fuel return hose. Who is correct?

A. A only

B. B only

C. Both A and B

D. Neither A nor B

Answer A is correct. Only Technician A is correct. Regular fuel hose cannot be used on a fuel injected vehicle. Fuel injection produces higher pressures than an old carbureted system, so the hose must be rated as fuel injection hose.

Answer B is incorrect. The fuel supply hose is usually the bigger hose of the two hoses, typically $\frac{5}{16}$-inch; the usual return hose is ¼-inch.

Answer C is incorrect. Only Technician A is correct.

Answer D is incorrect. Technician A is correct.

TASK D.1.2

49. Technician A says oil found inside the air filter housing may be caused by excessive crankcase blow-by. Technician B says a plugged PCV valve will cause excessive crankcase pressure. Who is correct?

A. A only

B. B only

C. Both A and B

D. Neither A nor B

Answer A is incorrect. Technician B is also correct.

Answer B is incorrect. Technician A is also correct.

Answer C is correct. Both Technicians are correct. A positive crankcase ventilation system vents the engine from excessive crankcase pressures. Excessive crankcase pressure can cause oil in the air cleaner, leaking seals, and leaking gaskets. If any of these conditions are present, then the PCV system should be inspected and tested.

Answer D is incorrect. Both Technicians are correct.

TASK E.2

50. A DTC for an intake air temperature sensor may be set by any of the following conditions EXCEPT:

A. An open in the voltage reference wire

B. Operating the vehicle in extremely cold climates

C. A short in the voltage reference wire

D. An out-of-range sensor input

Answer A is incorrect. An open in the voltage reference wire will cause a voltage high code.

Answer B is correct. Operating the vehicle in extremely cold weather will not cause a DTC to be set.

Answer C is incorrect. A short in the voltage reference wire will cause a voltage low code.

Answer D is incorrect. An out-of-range sensor will cause a DTC to be set.

PREPARATION EXAM 2—ANSWER KEY

1.	A	21.	A	41.	B
2.	D	22.	D	42.	D
3.	D	23.	A	43.	B
4.	B	24.	B	44.	C
5.	D	25.	C	45.	C
6.	D	26.	C	46.	B
7.	C	27.	A	47.	B
8.	A	28.	D	48.	A
9.	C	29.	A	49.	A
10.	C	30.	C	50.	C
11.	B	31.	A		
12.	C	32.	A		
13.	C	33.	C		
14.	B	34.	D		
15.	C	35.	C		
16.	D	36.	D		
17.	C	37.	D		
18.	A	38.	D		
19.	A	39.	C		
20.	C	40.	D		

PREPARATION EXAM 2—EXPLANATIONS

1. Which of the followings is the most likely cause of blue exhaust smoke?

    A. Worn piston rings
    B. Worn piston pins
    C. Rich fuel mixtures
    D. Internal coolant leak

TASK A.4

Answer A is correct. Worn piston rings allow oil to leak into the combustion chamber, where it burns, developing a blue-colored exhaust smoke.

Answer B is incorrect. Worn piston pins would cause a knock, not blue smoke.

Answer C is incorrect. A rich fuel mixture would cause black smoke.

Answer D is incorrect. An internal coolant leak would cause white smoke.

TASK C.5

2. All of the following components are part of the fuel pump control circuit EXCEPT:

    A. Fuel pump relay
    B. Powertrain control module (PCM)
    C. Ignition switch
    D. Body control module (BCM)

    Answer A is incorrect. The fuel pump relay sends power to the fuel pump.

    Answer B is incorrect. The PCM grounds the fuel pump relay to energize the fuel pump.

    Answer C is incorrect. The ignition switch supplies the power to the PCM and the fuel pump relay.

    Answer D is correct. The BCM controls body system devices, not the fuel pump.

TASK D.2.3

3. An EGR vacuum regulator solenoid (EGRV) is thought to be inoperative. Technician A says when measuring the resistance of the solenoid, a lower-than-specified reading means the windings are open. Technician B says an infinite reading means the winding is shorted. Who is correct?

    A. A only
    B. B only
    C. Both A and B
    D. Neither A nor B

    Answer A is incorrect. A reading lower than the specified indicates a shorted winding.

    Answer B is incorrect. An infinite reading means the windings are open.

    Answer C is incorrect. Neither Technician is correct.

    Answer D is correct. Neither Technician is correct. When diagnosing a solenoid, the solenoid is measured for resistance and compared to the specification. If the solenoid resistance is lower or higher than the specification, then replace the solenoid.

TASK E.8

4. Technician A says most computer inputs are received from other computers. Technician B says one input might affect other computers. Who is correct?

    A. A only
    B. B only
    C. Both A and B
    D. Neither A nor B

    Answer A is incorrect. While some computer inputs are received from other computers, most inputs are not.

    Answer B is correct. Only Technician B is correct. The PCM, BCM, and anti-lock brake module may all share information from one input sensor.

    Answer C is incorrect. Only Technician B is correct.

    Answer D is incorrect. Technician B is correct.

5. A cylinder-power balance test on a distributor-ignition vehicle can indicate all of the following problems EXCEPT:

   A.   Burnt valve
   B.   Cracked piston
   C.   Blown head gasket
   D.   Weak ignition coil

TASK A.6

   Answer A is incorrect. A burnt valve would cause a miss that could be located performing a power balance test.

   Answer B is incorrect. A cracked piston could cause a loss of compression that could lead to a miss. The problem could be found by performing a power balance test.

   Answer C is incorrect. A blown gasket could cause a miss due to the loss of compression, and it could be found by performing a power balance test.

   Answer D is correct. A weak coil would affect all the cylinders, so a power balance test would not help find it.

6. While testing the secondary ignition with an oscilloscope, which of the following is the LEAST LIKELY cause of high resistance in the ignition secondary circuit?

   A.   A burnt rotor button
   B.   Corroded ignition coil terminal
   C.   Wide spark plug gap
   D.   No dielectric grease on the ignition module mounting surface

TASK B.4

   Answer A is incorrect. The rotor button is part of the secondary and as the rotor button burns, the secondary resistance increases.

   Answer B is incorrect. Corroded ignition coil terminals would cause high secondary resistance.

   Answer C is incorrect. A wide spark plug gap increases secondary resistance.

   Answer D is correct. The lack of dielectric grease on the back of an ignition module will cause the module to overheat, but will not increase the secondary resistance.

7. Technician A says some manufacturers do not allow the cleaning of a throttle body with throttle body cleaner. Technician B says a buildup of gum and carbon deposits may cause rough idle operation. Who is correct?

   A.   A only
   B.   B only
   C.   Both A and B
   D.   Neither A nor B

TASK C.7

   Answer A is incorrect. Technician B is also correct.

   Answer B is incorrect. Technician A is also correct.

   Answer C is correct. Both Technicians are correct. The throttle body will begin to have a buildup of carbon over time on some throttle bodies. This causes the idle air control valve to open more to compensate. If the carbon buildup is bad enough, then a rough idle or stalling can occur. Some manufacturers to not allow the cleaning of throttle bodies—they require the replacement of the throttle body if it has carbon buildup.

   Answer D is incorrect. Both Technicians are correct.

TASK D.1.2

8. Technician A says with the PCV valve disconnected from the rocker cover, there should be vacuum at the valve with the engine idling. Technician B says when the PCV valve is removed and shaken, there should not be a rattling noise. Who is correct?

A. A only

B. B only

C. Both A and B

D. Neither A nor B

Answer A is correct. Only Technician A is correct. At idle with the PCV valve removed from the valve cover, there should be vacuum at the PCV valve.

Answer B is incorrect. A rattle test is another way to check to see if the plunger is free to move.

Answer C is incorrect. Only Technician A is correct.

Answer D is incorrect. Technician A is correct.

TASK E.1

9. Technician A says that if an emissions-related DTC is set, then a freeze frame is also stored. Technician B says if an emissions-related DTC is set, then the malfunction indicator lamp will illuminate. Who is correct?

A. A only

B. B only

C. Both A and B

D. Neither A nor B

Answer A is incorrect. Technician B is also correct.

Answer B is incorrect. Technician A is also correct.

Answer C is correct. Both Technicians are correct. The malfunction indicator lamp illuminates to warn the driver that the vehicle is malfunctioning and emitting emissions 1.5 times the federal test procedure allowance. When an emissions-related fault is set, a freeze frame of the scan tool data is stored in the PCM.

Answer D is incorrect. Both Technicians are correct.

TASK A.3

10. Which of the following is most likely to cause a double-knocking noise with the engine at an idle?

A. Excessive main bearing thrust clearance

B. Worn main bearing

C. Worn piston wrist pins

D. Excessive rod bearing clearance

Answer A is incorrect. Excessive main bearing thrust clearance will cause the crankshaft to have too much end-play, but will not cause a double-knock.

Answer B is incorrect. Worn main bearings will usually cause a deep low-end knock at idle and at start up, but will not cause a double-knock.

Answer C is correct. Worn piston wrist pins create a double-knock because the piston changes direction at TDC (top dead center) and BDC (bottom dead center) every crankshaft revolution.

Answer D is incorrect. Excessive rod bearing clearance will cause a knock that is more noticeable on acceleration.

11. Technician A says the plug wire resistance should not exceed 10 megohms of resistance. Technician B says plug wire insulation can be checked with a saltwater solution in a spray bottle. Who is correct?

    TASK B.4

    A. A only
    B. B only
    C. Both A and B
    D. Neither A nor B

    Answer A is incorrect. The general specification for spark plug wires is 5,000 to 10,000 ohms per foot.

    Answer B is correct. Only Technician B is correct. If the spark plug wire insulation breaks down, then the spark can ground at other areas besides the spark plug. To check for this problem, spray a saltwater solution along the spark plug wire. If an arc is found, then the spark plug wire should be replaced.

    Answer C is incorrect. Only Technician B is correct.

    Answer D is incorrect. Technician B is correct.

12. Technician A says a pressure drop test can be performed to test for a restricted fuel filter. Technician B says some vehicles do not use an in-line filter; the fuel filter is in the fuel tank. Who is correct?

    TASK C.4

    A. A only
    B. B only
    C. Both A and B
    D. Neither A nor B

    Answer A is incorrect. Technician B is also correct.

    Answer B is incorrect. Technician A is also correct.

    Answer C is correct. Both Technicians are correct. A pressure drop test is a very accurate way to check for a restricted fuel filter; however, some vehicles do not have a fuel filter inline to the fuel injector. Their fuel filter is in the fuel tank and is part of the fuel pump module.

    Answer D is incorrect. Both Technicians are correct.

13. Technician A says that *enabling criteria* are specific conditions that must be met before a monitor will run, such as coolant temperature or engine speed. Technician B says that *pending conditions* are conditions that exist that prevent a specific monitor from running, such as an oxygen sensor fault code preventing an oxygen sensor heater monitor from running. Who is correct?

    TASK E.1

    A. A only
    B. B only
    C. Both A and B
    D. Neither A nor B

    Answer A is incorrect. Technician B is also correct.

    Answer B is incorrect. Technician A is also correct.

    Answer C is correct. Both Technicians are correct. Enabling criteria are the specific operating parameters that must be met before a monitor will run. Many EVAP monitors, for instance, require a cold start with ambient temperature below a certain value. Pending conditions are any circumstances that may prevent a monitor from running properly, such as an oxygen sensor fault not allowing a catalyst monitor to run.

    Answer D is incorrect. Both Technicians are correct.

TASK A.1

14. Technician A says that the first step of any diagnostic procedure is to check for diagnostic trouble codes (DTCs). Technician B says the search for technical service bulletins (TSB) should be consulted before any repairs are made. Who is correct?

A. A only

B. B only

C. Both A and B

D. Neither A nor B

Answer A is incorrect. The first step in the diagnostic procedure is to verify the customer complaint.

Answer B is correct. Only Technician B is correct. Sometimes the fix is as simple as performing a procedure found in a technical service bulletin. If a vehicle has the same symptoms described in a TSB, then the TSB instructions should be performed before further repair is made.

Answer C is incorrect. Only Technician B is correct.

Answer D is incorrect. Technician B is correct.

TASK B.1

15. A primary ignition circuit on a vehicle checks good, but there is no spark from the spark plug wire. There is spark at the coil wire. This could be caused by any of the following:

A. Bad ignition coil

B. Bad ignition module

C. Shorted rotor button

D. Broken distributor gear

Answer A is incorrect. A bad coil would not allow spark at the coil wire ether.

Answer B is incorrect. A bad ignition module would not allow any spark.

Answer C is correct. If spark is going into the distributor but not coming out, then the rotor button should be checked.

Answer D is correct. A broken distributor gear would not allow any spark.

TASK C.14

16. Technician A says the wastegate on a turbocharger closes to control boost by redirecting exhaust gases around the turbine. Technician B says once boost pressure is under control, the wastegate opens completely. Who is correct?

A. A only

B. B only

C. Both A and B

D. Neither A nor B

Answer A is incorrect. The wastegate opens to control boost, not close.

Answer B is incorrect. If an over-boost condition does not exist, then the wastegate closes completely.

Answer C is incorrect. Neither Technician is correct.

Answer D is correct. Neither Technician is correct. The wastegate is used to control the boost of a turbocharger. If the boost pressure exceeds the wastegate diaphragm rating, then the wastegate will open to lower the boost. Once the boost is under control, the wastegate will close.

17. All of the following are part of the secondary air system EXCEPT:

   A. Exhaust check valve
   B. Diverter valve
   C. Back pressure transducer
   D. Air pump

TASK D.3.3

Answer A is incorrect. The exhaust check valve prevents the flow of exhaust in to the secondary air components.

Answer B is incorrect. The diverter valve diverts the secondary air upstream, downstream, or to the air cleaner.

Answer C is correct. The back pressure transducer is part of the exhaust gas recirculation system.

Answer D is incorrect. The air pump supplies additional air to the catalytic converter.

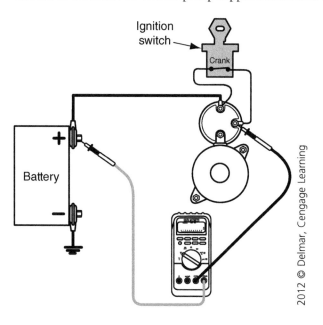

18. Technician A says an ignition switch starter-control circuit voltage drop test is being performed in the above illustration. Technician B says a reading of 2 volts is acceptable on this circuit. Who is correct?

TASK E.5

   A. A only
   B. B only
   C. Both A and B
   D. Neither A nor B

Answer A is correct. Only Technician A is correct. A voltage drop test is being performed on the starter control circuit.

Answer B is incorrect. The maximum allowable voltage drop on the starter-control circuit is 0.3 volts; a 2-volt drop is excessive.

Answer C is incorrect. Only Technician A is correct.

Answer D is incorrect. Technician A is correct.

TASK A.7

19. Technician A says that during a cylinder compression test, low readings on adjacent cylinders may be caused by a cracked cylinder head. Technician B says a low reading on a single cylinder that increases when a tablespoon of oil is added is probably a valve problem. Who is correct?

A. A only

B. B only

C. Both A and B

D. Neither A nor B

Answer A is correct. Only Technician A is correct. Any time adjacent cylinders are both low, a blown head gasket or cracked head should be suspected.

Answer B is incorrect. If a cylinder reads low compression but increases with the addition of a little oil in the cylinder, then the most likely problem is in the rings, not the valves.

Answer C is incorrect. Only Technician A is correct.

Answer D is incorrect. Technician A is correct.

TASK B.5

20. Technician A says when testing the ignition coil, both the primary and secondary winding should be checked for resistance. Technician B says maximum coil output testing can be performed with an oscilloscope. Who is correct?

A. A only

B. B only

C. Both A and B

D. Neither A nor B

Answer A is incorrect. Technician B is also correct.

Answer B is incorrect. Technician A is also correct.

Answer C is correct. Both Technicians are correct. When testing the ignition coil, the coil resistance of the primary and secondary windings should be compared with the manufacturer's specifications. The primary winding should measure low, around 1 to 2 ohms. The secondary winding should measure around 8,000 to 12,000 ohms. Another test that can be performed is a maximum coil output test; this can be performed using an oscilloscope.

Answer D is incorrect. Both Technicians are correct.

TASK C.14

21. A seized-up turbocharger can cause vacuum readings to do which of the following?

A. Show a continuous gradual drop as engine speed is increased

B. Fluctuate between 15 and 21 inches at an idle

C. Drop off about 3 inches at an idle

D. Drop off about 6 inches at an idle

Answer A is correct. A seized-up turbocharger can create a restricted exhaust which will cause the vacuum to continuously and gradually drop as the engine speed is increased.

Answer B is incorrect. A fluctuating needle at idle usually indicates a burn valve.

Answer C is incorrect. A drop in vacuum at idle or low vacuum reading can be caused by a vacuum leak or late valve timing.

Answer D is incorrect. A drop in vacuum at idle or low vacuum reading can be caused by a vacuum leak or late valve timing.

22. Which of the follow DTCs are the highest priority codes and should be diagnosed first?

    A. Transmission trouble codes
    B. Emissions-related trouble codes
    C. Fuel-related trouble codes
    D. Misfire-related trouble codes

TASK E.1

    Answer A is incorrect. Transmission codes are not high-priority codes.

    Answer B is incorrect. Emissions-related codes will turn the MIL (malfunction indicator lamp) on, but are not the highest-priority code listed.

    Answer C is incorrect. Fuel-related codes are not the highest-priority codes listed.

    Answer D is correct. When a misfire occurs, the catalytic converter can be damaged; thus, misfire codes are of the highest priority and should be diagnosed first.

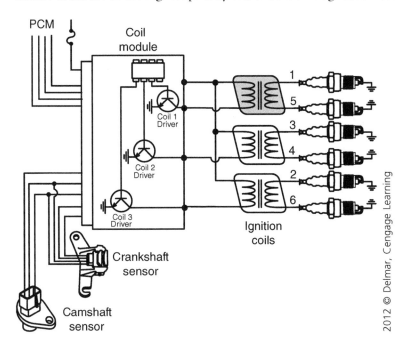

23. A vehicle with the ignition system illustrated has a no-spark condition on cylinders #1 and #5. Technician A says the ignition module could have a bad coil driver. Technician B says the crankshaft sensor could have an open circuit, causing the no spark at #1 and #5. Who is correct?

TASK B.1

    A. A only
    B. B only
    C. Both A and B
    D. Neither A nor B

    Answer A is correct. Only Technician A is correct. If the ignition control module were to lose a coil driver for ignition coil pack #1, then cylinder #1 and #5 would not have any spark.

    Answer B is incorrect. If the crank sensor were to become open, then the vehicle would have no spark at any of the coils, not just one coil pack.

    Answer C is incorrect. Only Technician A is correct.

    Answer D is incorrect. Technician A is correct.

TASK A.18

24. All of the following could cause high starter current readings EXCEPT:

    A. A short circuit

    B. An open circuit

    C. Tight engine crankshaft bearings

    D. A defective starter

    Answer A is incorrect. A shorted circuit in the starter circuit could cause the starter to read a high current draw.

    Answer B is correct. An open circuit would not allow any current flow.

    Answer C is incorrect. If the crankshaft bearings are too tight, then the starter current draw will be too high.

    Answer D is incorrect. A defective starter can cause the starter current draw to be too high.

TASK C.6

25. A vehicle with sequential fuel injection (SFI) has high fuel pump pressure at idle. Which of the following could be the cause?

    A. Excessive manifold vacuum

    B. A leaking fuel pump drain-back check valve

    C. Low manifold vacuum

    D. A stuck-closed fuel injector

    Answer A is incorrect. Excessive manifold vacuum would have no effect on fuel pump pressure.

    Answer B is incorrect. The fuel pump check valve is used to keep residual pressure in the fuel lines when the engine is off. A leaking check valve would cause a hard-starting complaint, but would not affect fuel pressure.

    Answer C is correct. Manifold vacuum is used by the fuel regulator to control fuel pressure; as the manifold vacuum drops, the fuel pressure increases. With the vacuum hose removed from the fuel regulator, the fuel pressure should increase about 10 psi.

    Answer D is incorrect. A stuck-closed fuel injector would cause an engine miss, but would not affect the fuel pressure.

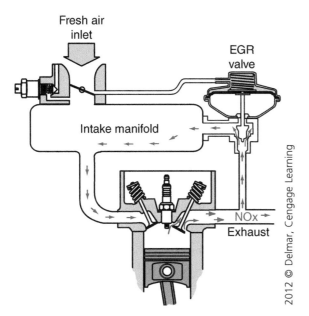

Fresh air inlet

EGR valve

Intake manifold

NOx

Exhaust

2012 © Delmar, Cengage Learning

26. Technician A says the exhaust gas recirculation valve (EGR) illustrated allows engine exhaust to enter the intake manifold of the engine. Technician B says the EGR system lowers combustion chamber temperature. Who is correct?

A. A only
B. B only
C. Both A and B
D. Neither A nor B

TASK D.2.3

Answer A is incorrect. Technician B is also correct.

Answer B is incorrect. Technician A is also correct.

Answer C is correct. Both Technicians are correct. The EGR system is used to reduce the formation of oxides of nitrogen in the engine. It redirects exhaust gases back into the intake manifold. These gases take up room in the combustion chamber that would otherwise be filled with fuel and air. By reducing the amount of air/fuel the combustion chamber can accept, the combustion chamber temperature is lowered.

Answer D is incorrect. Both Technicians are correct.

27. Technician A says a scan tool or code reader is used to retrieve DTCs from an OBD II vehicle. Technician B says the DTCs can be retrieved by watching and counting the number of flashes from the MIL on OBD (on-board diagnostics) II vehicle. Who is correct?

A. A only
B. B only
C. Both A and B
D. Neither A nor B

TASK E.1

Answer A is correct. Only Technician A is correct. A scan tool or code reader must be used to retrieve DTCs from an OBD II vehicle.

Answer B is incorrect. On OBD I vehicles, the DTCs could be read by counting the flashes of the MIL, but OBD II vehicles require the use of a scan tool or code reader.

Answer C is incorrect. Only Technician A is correct.

Answer D is incorrect. Technician A is correct.

TASK A.9

28. A multi-trace oscilloscope can test all of the following EXCEPT:

    A. Throttle position sensor (TPS)

    B. Mass air flow sensor (MAF)

    C. Crankshaft position (CKP) sensor

    D. Air/fuel ratio

    Answer A is incorrect. The TPS can be tested using an oscilloscope.

    Answer B is incorrect. The oscilloscope is a good choice of tools to use to test a MAF.

    Answer C is incorrect. If the CKP sensor is suspected of being the cause of a no-start condition, then an oscilloscope can be used to check for a CKP sensor signal.

    Answer D is correct. The air/fuel ratio cannot be tested using an oscilloscope. A 4- or 5-gas analyzer would be used for this task.

TASK B.4

29. Technician A says the secondary ignition circuit is designed to handle high voltages up in the thousand volts. Technician B says the secondary circuit produces dangerously high amperage. Who is correct?

    A. A only

    B. B only

    C. Both A and B

    D. Neither A nor B

    Answer A is correct. Only Technician A is correct. The secondary ignition system is designed to handle very high voltages, some as high as 70,000 or 80,000 volts.

    Answer B is incorrect. Although the secondary ignition circuit produces high voltage, the amperage produced is very low and is not enough to be harmful. Many systems also include current-limiting circuitry in the module to control this problem.

    Answer C is incorrect. Only Technician A is correct.

    Answer D is incorrect. Technician A is correct.

TASK C.13

30. A vehicle has a low power complaint with a hissing sound coming from under the vehicle at wide-open throttle (WOT). Technician A says the exhaust back pressure should be checked. Technician B says the manifold vacuum should be checked. Who is correct?

    A. A only

    B. B only

    C. Both A and B

    D. Neither A nor B

    Answer A is incorrect. Technician B is also correct.

    Answer B is incorrect. Technician A is also correct.

    Answer C is correct. Both Technicians are correct. The low power with a hissing noise is a good indication of a restricted exhaust. A back-pressure gauge or a vacuum gauge can be used to check for a restricted exhaust. If the back pressure exceeds 2.5 psi at 2,500 rpm or the vacuum drops and continues to drop at 2,500 rpm, then a restricted exhaust is indicated.

    Answer D is incorrect. Both Technicians are correct.

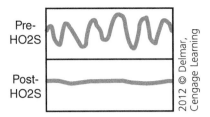

Pre-HO2S

Post-HO2S

2012 © Delmar, Cengage Learning

31. A multi-trace oscilloscope is being used on the pre- and post-catalytic converter heated oxygen sensors, and the results are shown as illustrated. Technician A says the catalytic converter is working correctly. Technician B says the post-converter heated O_2 (oxygen) sensor should be replaced. Who is correct?

TASK D.3.4

   A. A only

   B. B Only

   C. Both A and B

   D. Neither A nor B

Answer A is correct. Only Technician A is correct. If the catalytic converter (CAT) is working correctly, then the pre-CAT heated oxygen sensor will have normal switch between 900 millivolts and 100 millivolts just as illustrated.

Answer B is incorrect. The post-CAT heated oxygen sensor is working just like it should. If the CAT was not working, then the post-CAT heated oxygen sensor would look just like the pre-CAT heated oxygen sensor.

Answer C is incorrect. Only Technician A is correct.

Answer D is incorrect. Technician A is correct.

32. Which of the following conditions is most likely caused by a manifold absolute pressure (MAP) sensor?

TASK E.4

   A. Poor fuel economy

   B. Excessive idle speeds

   C. Spark knock

   D. Erratic speedometer operation

Answer A is correct. The MAP sensor is used for load detection; a bad MAP sensor can definitely cause poor fuel economy complaints.

Answer B is incorrect. Excessive idle speeds are usually traced back to the idle air control circuit or the throttle position sensor circuit, not the MAP sensor.

Answer C is incorrect. Spark knock is caused by over advanced timing or an overheated combustion chamber, not a defective MAP sensor.

Answer D is incorrect. Erratic speedometer operation is usually traced to the vehicle speed sensor circuit, not the MAP sensor.

TASK A.15

33. What behavior should a viscous-drive fan clutch exhibit when rotated by hand with the engine off?

   A. No resistance

   B. More resistance cold

   C. More resistance hot

   D. Not freewheel any

Answer A is incorrect. The fan clutch will have resistance; the amount will vary with temperature.

Answer B is incorrect. A cold engine does not require the full operation of a cooling fan, so the fan will have less resistance when cold.

Answer C is correct. A hot engine requires a lot of air flow through the radiator for cooling, especially at idle. The hotter the engine, the more the resistance will get in the viscous-drive fan clutch.

Answer D is incorrect. If the fan clutch will not freewheel, then it is locked up and must be replaced.

TASK B.3

34. When installing and timing the distributor, Technician A says the engine must be timed referencing TDC on the specified cylinder's exhaust stroke. Technician says if the engine is timed on the compression stroke, then the distributor will be 180 degrees off. Who is correct?

   A. A only

   B. B only

   C. Both A and B

   D. Neither A nor B

Answer A is incorrect. The distributor is timed to cylinder #1 on the compression stroke.

Answer B is incorrect. The distributor is timed to cylinder #1 on the compression stroke.

Answer C is incorrect. Neither Technician is correct.

Answer D is correct. Neither Technician is correct. When timing the distributor, piston #1 is brought to TDC on the compression stroke, then the distributor is installed. Once started, the ignition timing must be set using a timing light.

TASK C.13

35. Technician A says restricted exhaust may cause reduced fuel economy. Technician B says restricted exhaust causes reduced engine efficiency. Who is correct?

   A. A only

   B. B only

   C. Both A and B

   D. Neither A nor B

Answer A is incorrect. Technician B is also correct.

Answer B is incorrect. Technician A is also correct.

Answer C is correct. Both Technicians are correct. A restricted exhaust prevents the engine from drawing in the needed air/fuel mixture because it cannot get rid of the inert exhaust fast enough. Restricted exhaust lowers power, fuel economy, and engine efficiency.

Answer D is incorrect. Both Technicians are correct.

36. A hesitation during acceleration from a stop on a fuel-injected engine may be caused by all of the following EXCEPT:

    A. A purge-control solenoid
    B. A faulty MAP sensor
    C. A faulty TPS
    D. A faulty vehicle-speed sensor (VSS)

TASK E.4

Answer A is incorrect. If the purge-control solenoid allows canister purge at the wrong time, then the vehicle will hesitate on acceleration.

Answer B is incorrect. A faulty MAP sensor will cause the fuel mixture to be incorrect and can cause a hesitation on acceleration.

Answer C is incorrect. If the TPS gets a bad spot on it, then it can cause a hesitation.

Answer D is correct. A faulty speed sensor will affect the electronic engine control system, but not in a way that would cause a hesitation.

37. An engine has a lack of power and excessive fuel consumption. Technician A says a broken timing belt could be the cause. Technician B says the timing belt may have jumped a tooth 180 degrees out of time. Who is correct?

    A. A only
    B. B only
    C. Both A and B
    D. Neither A nor B

TASK A.12

Answer A is incorrect. A broken timing belt would cause a no-start condition.

Answer B is incorrect. If the timing belt were to jump 180 degrees out of time, then the engine would not start at all.

Answer C is incorrect. Neither Technician is correct.

Answer D is correct. Neither Technician is correct. A timing belt that was one tooth off could cause this complaint; however, a broken or jumped timing belt 180 degrees would cause a no-start.

38. What is the purpose of coating the back of an ignition module with heat sink grease (sometimes referred to as dielectric silicone grease) before installation?

    A. To insulate the module from excessive voltage spikes
    B. To electrically insulate the module from ground
    C. To ensure electrical ground
    D. To help dissipate heat from the module

TASK B.7

Answer A is incorrect. Dielectric grease is not used to suppress voltage spikes.

Answer B is incorrect. The module needs to be grounded; dielectric grease does not insulate the module from ground.

Answer C is incorrect. The ignition module is grounded externally or through a mounting bolt.

Answer D is correct. The dielectric grease helps transfer the heat generated in the ignition module away from the module.

TASK C.9

39. Technician A says a damaged or missing air filter can increase wear on cylinder walls. Technician B says an air filter problem can affect fuel consumption. Who is correct?

   A.   A only

   B.   B only

   C.   Both A and B

   D.   Neither A nor B

   Answer A is incorrect. Technician B is also correct.

   Answer B is incorrect. Technician A is also correct.

   Answer C is correct. Both Technicians are correct. A damaged or missing air filter will allow dirt particles to enter the combustion chamber and cause abrasive wear to the cylinder walls. If the air filter gets dirty, then it becomes harder and harder for the air to enter the intake manifold. This causes the air/fuel mixture to become rich, causing excessive fuel consumption and drivability complaints.

   Answer D is incorrect. Both Technicians are correct.

TASK D.4.2

40. An evaporative system DTC may be set by all of the following EXCEPT:

   A.   A cracked vacuum hose

   B.   A loose gas cap

   C.   An open purge control solenoid

   D.   A leaking fuel injector

   Answer A is incorrect. A cracked vacuum hose will cause an EVAP leak code.

   Answer B is incorrect. A loose gas cap will cause a large EVAP leak code. Some manufacturers have a check gas cap light on the dash.

   Answer C is incorrect. An open purge solenoid would cause a DTC to be set.

   Answer D is correct. While the EVAP system prevents hydrocarbons (HCs) from entering the atmosphere, it cannot detect a leaking fuel injector.

TASK E.7

41. Technician A says the PCM cannot be harmed with static electricity if the negative and positive battery cables are disconnected. Technician B says you should always ground yourself to the vehicle while working on a PCM. Who is correct?

   A.   A only

   B.   B only

   C.   Both A and B

   D.   Neither A nor B

   Answer A is incorrect. Disconnecting the positive and negative battery cable will not insure that a static charge of electricity will not be transferred from the technician to the PCM.

   Answer B is correct. Only Technician B is correct. It is always good practice to ground yourself with the vehicle before removing the PCM, this can be done using a special wrist grounding strap.

   Answer C is incorrect. Only Technician B is correct.

   Answer D is incorrect. Technician B is correct.

42. All of the following are part of the diagnostic process EXCEPT:

TASK A.1

   A. Verify the complaint

   B. Road test the vehicle

   C. Perform a visual inspection

   D. Re-flash the PCM

   Answer A is incorrect. The complaint should always be verified before any work is performed.

   Answer B is incorrect. The vehicle should be road tested under the same conditions as when the complaint occurs.

   Answer C is incorrect. Many drivability complaints are fixed by performing a visual inspection.

   Answer D is correct. Re-flashing the PCM may be needed to repair a drivability complaint, but it is not part of the diagnostic process.

43. A sequential fuel-injected vehicle has a rough idle. Technician A says this could be caused by a cracked hose between the fuel tank and the EVAP canister. Technician B says a malfunctioning EVAP purge solenoid can cause idle problems. Who is correct?

TASK D.4.3

   A. A only

   B. B only

   C. Both A and B

   D. Neither A nor B

   Answer A is incorrect. A cracked fuel tank vent hose would cause fuel smell and an EVAP DTC to be set, but would not cause a drivability complaint.

   Answer B is correct. Only Technician B is correct. If an EVAP purge solenoid were to allow purging of the canister at idle, then the vehicle could idle rough and hesitate when accelerating from a stop.

   Answer C is incorrect. Only Technician B is correct.

   Answer D is incorrect. Technician B is correct.

44. While scanning an OBD II vehicle for DTCs, a P1000 is retrieved. Technician A says that a first digit of P means the code is a powertrain trouble code. Technician B says that a second digit of 1 means the code is a manufacturer-specific code. Who is correct?

TASK E.1

   A. A only

   B. B only

   C. Both A and B

   D. Neither A nor B

   Answer A is incorrect. Technician B is also correct.

   Answer B is incorrect. Technician A is also correct.

   Answer C is correct. Both Technicians are correct. The first digit of a code indicates what type of code it is (P for powertrain, B for body, U for network, and C for chassis). The second digit indicates whether the code is generic or manufacturer-specific (1 for manufacturer-specific and 0 for generic).

   Answer D is incorrect. Both Technicians are correct.

TASK A.3

45. Technician A says that worn valve train components usually produce a clicking noise. Technician B says an engine noise diagnosis should be performed before doing engine repair work. Who is correct?

A. A only
B. B only
C. Both A and B
D. Neither A nor B

Answer A is incorrect. Technician B is also correct.

Answer B is incorrect. Technician A is also correct.

Answer C is correct. Both Technicians are correct. Before any engine repair work is performed, the cause of the noise must be identified. A clicking noise is usually in the valve train. A double-knocking noise is usually a piston wrist pin. A low-sounding knock heard on startup is usually a main bearing. A rod bearing produces a knock on acceleration.

Answer D is incorrect. Both Technicians are correct.

TASK D.4.1

46. Which of the following is the most likely symptom resulting from an evaporative emissions system failure?

A. Increased tail pipe emissions
B. Fuel odor
C. Engine miss at highway speed
D. Hard to start on a cold engine

Answer A is incorrect. An EVAP system failure would allow fuel vapors to escape into the atmosphere. When the purge solenoid failed, the vehicle might hesitate, but tail pipe emissions are not the most likely symptom.

Answer B is correct. The release of fuel vapors would produce a fuel odor.

Answer C is incorrect. A failure in the purge solenoid could cause a hesitation or a rough idle, but engine miss at highway speeds is not the most likely.

Answer D is incorrect. A failure of the EVAP system will not cause a hard-to-start-cold symptom.

TASK E.7

47. A vehicle needs to have its PCM re-flashed. Technician A says the vehicle should be left running to prevent accidental battery discharge during flashing. Technician B says the scan tool must fit in the DLC (diagnostic link connector) snuggly. Who is correct?

A. A only
B. B only
C. Both A and B
D. Neither A nor B

Answer A is incorrect. The vehicle should be connected to a battery charger to prevent the battery from discharging during flashing, not left running.

Answer B is correct. Only Technician B is correct. When re-flashing a PCM, the old program is erased and a new one installed. If the battery runs down or the scan tool comes loose from the DLC, then all information could be lost to the point that communication with the PCM cannot be reestablished and a new PCM must be purchased.

Answer C is incorrect. Only Technician B is correct.

Answer D is incorrect. Technician B is correct.

48. When diagnosing a fuel injection system problem, a technical service bulletin search is performed for all of the following reasons EXCEPT:

TASK A.2

   A. Year, make, and vehicle identification number
   B. Midyear production changes
   C. Service manual updated
   D. View revisions to existing procedures

Answer A is correct. The year, make and vehicle identification number should already be known before doing a TSB search.

Answer B is incorrect. When a midyear production change is made, a technical service bulletin is issued.

Answer C is incorrect. If a mistake is found in a service manual, then a technical service bulletin is issued.

Answer D is incorrect. When revisions are made to existing procedures a technical service bulletin is issued.

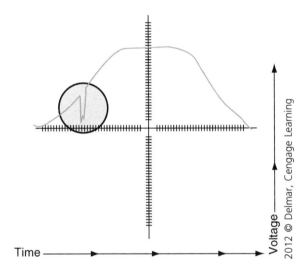

49. Technician A says the above illustration shows a defective TPS. Technician B says it is a scope pattern of the MAP sensor. Who is correct?

TASK E.4

   A. A only
   B. B only
   C. Both A and B
   D. Neither A nor B

Answer A is correct. Only Technician A is correct. The pattern is of a TPS. The pattern shows a bad spot in the sensor.

Answer B is incorrect. The pattern is not of a MAP sensor, it is of a TPS.

Answer C is incorrect. Only Technician A is correct.

Answer D is incorrect. Technician A is correct.

TASK E.4

50. Technician A says a defective MAP sensor may cause a lean air/fuel ratio. Technician B says the MAP sensor can cause a no-start. Who is correct?

A. A only

B. B only

C. Both A and B

D. Neither A nor B

Answer A is incorrect. Technician B is also correct.

Answer B is incorrect. Technician A is also correct.

Answer C is correct. Both Technicians are correct. The MAP sensor measures engine load on a speed density system and relays this information to the PCM for injector pulse width and fuel trim corrections. On some occasions, the MAP sensor can cause a no-start.

Answer D is incorrect. Both Technicians are correct.

PREPARATION EXAM 3—ANSWER KEY

1.	D	21.	C	41.	C
2.	C	22.	D	42.	C
3.	A	23.	D	43.	D
4.	A	24.	B	44.	A
5.	C	25.	C	45.	A
6.	D	26.	C	46.	A
7.	C	27.	A	47.	D
8.	B	28.	B	48.	D
9.	A	29.	A	49.	C
10.	C	30.	C	50.	A
11.	C	31.	B		
12.	C	32.	C		
13.	B	33.	B		
14.	D	34.	A		
15.	A	35.	C		
16.	C	36.	D		
17.	D	37.	D		
18.	D	38.	C		
19.	C	39.	A		
20.	B	40.	C		

PREPARATION EXAM 3—EXPLANATIONS

1. All of the following are tests performed on the PCV system EXCEPT:

    A. The rattle test

    B. The snap-back test

    C. Crankcase vacuum test

    D. Blowby test

    TASK D.1.2

    Answer A is incorrect. The PCV valve should rattle when shaken back and forth.

    Answer B is incorrect. When the thumb is placed over the end of the PCV valve with the engine idling, the plunger should snap back.

    Answer C is incorrect. A water manometer can be used to check the crankcase for vacuum at idle.

    Answer D is correct. There is no blowby test that is performed on the PCV system.

2. Which of the following is the LEAST LIKELY cause of ignition module failure?

    A. An open spark plug wire

    B. No dielectric grease under the module

    C. A fouled spark plug

    D. Loose module mounting screws

    TASK B.7

    Answer A is incorrect. Faulty secondary ignition components are common causes of ignition module failure.

    Answer B is incorrect. Some modules require the use of dielectric grease to aid in dissipating heat.

    Answer C is correct. A fouled spark plug would cause an engine miss, but would not cause the ignition module to fail.

    Answer D is incorrect. Loose module mounting screws will cause the module to fail prematurely.

3. A sequential fuel-injected vehicle has poor fuel economy, yet starts and runs fine. Technician A says the fuel return line may be partially restricted. Technician B says the fuel pressure regulator may be stuck open. Who is correct?

    A. A only

    B. B only

    C. Both A and B

    D. Neither A nor B

    TASK C.6

    Answer A is correct. Only Technician A is correct. A partially restricted fuel return hose would cause the fuel pressure to be higher than it should. This would cause poor fuel economy without affecting drivability.

    Answer B is incorrect. A stuck open fuel regulator would lower fuel pressure, possibly causing a no-start, but would not cause a fuel economy complaint.

    Answer C is incorrect. Only Technician A is correct.

    Answer D is incorrect. Technician A is correct.

TASK D.2.3

4. Technician A says if the passages in the exhaust gas recirculation (EGR) system get plugged up with carbon, then the engine could spark knock. Technician B says if the EGR passages get plugged up, then the engine will not run hot enough for complete combustion. Who is correct?

A. A only
B. B only
C. Both A and B
D. Neither A nor B

Answer A is correct. Only Technician A is correct. The EGR system is used to lower combustion chamber temperature. If the EGR system does not operate, then the engine will operate too hot, causing spark knock and an elevated oxides of nitrogen (NOx) reading.

Answer B is incorrect. An inoperative EGR system will raise combustion chamber temperatures and cause the engine to run hotter.

Answer C is incorrect. Only Technician A is correct.

Answer D is incorrect. Technician A is correct.

TASK B.4

5. Technician A says a coil with weak reserve voltage could cause a miss under acceleration. Technician B says the ignition coil only produces enough secondary voltage to jump the spark plug gap. Who is correct?

A. A only
B. B only
C. Both A and B
D. Neither A nor B

Answer A is incorrect. Technician B is also correct.

Answer B is incorrect. Technician A is also correct.

Answer C is correct. Both Technicians are correct. When the power transistor in the ignition control module opens the primary ignition circuit, the built-up magnetic field in the coil collapses. This produces the secondary voltage to jump the gap at the spark plug. The voltage that the coil produces can be broken down into three different types: *required voltage* (the voltage required to jump the gap at the spark plug), *available voltage* (the voltage available to jump the gap at the spark plug), and the *reserve voltage* (the voltage in reserve after the spark plug fires). As the engine accelerates, the required voltage goes up. If there is not enough voltage in reserve, then the engine will miss.

Answer D is incorrect. Both Technicians are correct.

TASK A.1

6. Which of the following is the very first step a technician should take when performing a diagnostic procedure?

A. Perform simple tests
B. Retrieve diagnostic trouble codes (DTCs)
C. Check for technical service bulletins (TSBs)
D. Verify the customer complaint

Answer A is incorrect. Simple test should be performed, but not until the problem is verified.

Answer B is incorrect. The DTCs will eventually be retrieved, but not before verifying the complaint.

Answer C is incorrect. Once any trouble codes are retrieved, a technical service bulletin search should be performed, but not before the complaint is verified.

Answer D is correct. The very first step to doing any diagnostic procedure is to verify the customer complaint.

7. All of the following could cause a multiple-cylinder misfire code EXCEPT:

    A. Low fuel pressure

    B. A leaking EGR valve

    C. A burnt valve

    D. Retarded valve timing

TASK B.2

Answer A is incorrect. Low fuel pressure would affect all the cylinders.

Answer B is incorrect. A leaking EGR valve would cause a misfire on multiple cylinders.

Answer C is correct. A burnt valve would only affect the cylinder with the burnt valve.

Answer D is incorrect. If the valve timing were retarded, then it would affect multiple cylinders.

8. A vehicle is equipped with a vented gas cap. Technician A says that if a non-vented cap is installed, then the vehicle could run rich at high speeds. Technician B says if a non-vented cap is installed in this vehicle, then the gas tank could collapse. Who is correct?

    A. A only

    B. B only

    C. Both A and B

    D. Neither A nor B

TASK C.3

Answer A is incorrect. If a non-vented gas cap were used in place of a vented cap, then the vehicle could starve for fuel due to the vacuum locking that would take place in the fuel tank.

Answer B is correct. Only Technician B is correct. As fuel cools down, the pressure in the fuel tank will drop. If a vented cap is not used, then the fuel tank pressure could drop low enough to cause a vacuum in the fuel tank; if the vacuum is strong enough, then the tank could collapse.

Answer C is incorrect. Only Technician B is correct.

Answer D is incorrect. Technician B is correct.

9. Technician A says secondary air injection is added to some vehicles to help oxidize hydrocarbon (HC) and carbon monoxide (CO) emissions. Technician B says secondary air is needed to lean out the air/fuel mixture on some vehicles. Who is correct?

    A. A only

    B. B only

    C. Both A and B

    D. Neither A nor B

TASK D.3.3

Answer A is correct. Only Technician A is correct. The purpose of the secondary air system is to inject additional air into the catalytic converter to help the converter oxidize the HC and CO emissions.

Answer B is incorrect. The secondary air system is never used to lean out the air/fuel mixture.

Answer C is incorrect. Only Technician A is correct.

Answer D is incorrect. Technician A is correct.

TASK E.1

10. Technician A says a scan tool is required to retrieve trouble codes on a vehicle with On-Board Diagnostics Second Generation (OBD II). Technician B says OBD II-compliant vehicles use a standardized trouble code format. Who is correct?

A.　A only

B.　B only

C.　Both A and B

D.　Neither A nor B

Answer A is incorrect. Technician B is also correct.

Answer B is incorrect. Technician A is also correct.

Answer C is correct. Both Technicians are correct. While codes could be retrieved without a scan tool on an OBD I system, a scan tool is required if the vehicle is an OBD II vehicle. Another feature of OBD II is a standardized trouble code format.

Answer D is incorrect. Both Technicians are correct.

TASK A.7

11. A Technician is performing a running compression test on a vehicle with suspected worn ring and cylinder problems. Technician A says running compression should be half of static compression at idle. Technician B says during a running compression test, the technician should increase the engine speed to 2,000 rpm, and the running compression should be lower than at idle. Who is correct?

A.　A only

B.　B only

C.　Both A and B

D.　Neither A nor B

Answer A is incorrect. Technician B is also correct.

Answer B is incorrect. Technician A is also correct.

Answer C is correct. Both Technicians are correct. Compression reading at idle should be approximately half of the cranking compression or typically between 60 and 90 psi. Compression reading at 2,000 rpm should be typically between 30 and 60 psi. As with cranking compression, the running compression of all cylinders should be equal.

Answer D is incorrect. Both Technicians are correct.

TASK B.5

12. Which of the following is LEAST LIKELY to be a test that would be performed on an ignition coil?

A.　Resistance primary to primary

B.　Resistance primary to secondary

C.　Secondary circuit amperage

D.　Primary circuit amperage

Answer A is incorrect. The resistance value of the primary circuit in the coil is crucial for good secondary circuit operation.

Answer B is incorrect. A resistance measurement of the primary to secondary measures the secondary winding resistance and is an important test.

Answer C is correct. The secondary ignition system is a high-voltage, low-amperage circuit. While the secondary voltage readings are important, the amperage is not measured.

Answer D is incorrect. The primary amperage is very important. If the primary amperage was low, then the secondary voltage would be low.

13. A vehicle with low engine power is being diagnosed. Technician A says an exhaust back-pressure test should be at least 5 psi at 2,000 rpm. Technician B says the catalytic converter may have come apart and be restricting exhaust flow. Who is correct?

TASK C.13

   A. A only
   B. B only
   C. Both A and B
   D. Neither A nor B

   Answer A is incorrect. A reading of 5 psi back pressure at 2,000 rpm is excessive and would indicate an exhaust restriction. Exhaust back pressure should not exceed 2.5 psi at 2,500 rpm.

   Answer B is correct. Only Technician B is correct. The catalytic converter can come apart causing an exhaust restriction. Sometimes this can be found by performing a converter rattle test.

   Answer C is incorrect. Only Technician B is correct.

   Answer D is incorrect. Technician B is correct.

14. Technician A says the filter on the engine-off natural vacuum pump (EONV) should be serviced with the regular oil change interval. Technician B says the EONV EVAP system does not use a vent valve. Who is correct?

TASK D.4.3

   A. A only
   B. B only
   C. Both A and B
   D. Neither A nor B

   Answer A is incorrect. The EONV system does not use a pump; instead, it closes the vent valve and monitors the fuel tank pressure as the fuel cools off. If the EVAP system is not leaking, then a specific amount of vacuum will be present on cool down.

   Answer B is incorrect. The vent valve is a naturally open valve that is closed by the PCM (powertrain control module) when it runs an EVAP monitor.

   Answer C is incorrect. Neither Technician is correct.

   Answer D is correct. Neither Technician is correct. Some systems use a leak detection pump to monitor the EVAP for leaks, but the EONV system uses the natural physics of heat and pressure.

15. All of the following apply to OBD II vehicles EXCEPT:

TASK E.2

   A. A standardized 26-pin data link connector (DLC)
   B. A standardized list of diagnostic trouble codes (DTCs)
   C. The ability to perform rationality tests on components
   D. A standardized communication protocol

   Answer A is correct. OBD II requires the use of a standardized 1-pin DLC, not 26-pin.

   Answer B is incorrect. ODB II requires the use of a standardized list of DTCs.

   Answer C is incorrect. OBD II vehicles can perform rationality tests on components.

   Answer D is incorrect. OBD II requires the use of a standardized communication protocol.

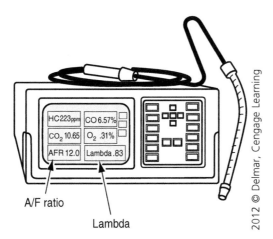

A/F ratio

Lambda

2012 © Delmar, Cengage Learning

TASK A.10

16. Refer to the gas analyzer readings illustration. Technician A says the vehicle is running too rich. Technician B says the high HC reading is from incomplete combustion. Who is correct?

   A. A only

   B. B only

   C. Both A and B

   D. Neither A nor B

Answer A is incorrect. Technician B is also correct. The ideal air/fuel ratio is 14.7:1. Anything above this level is lean, and anything below this is rich. The analyzer shows 12 for air/fuel ratio, which is rich.

Answer B is incorrect. Technician A is also correct. The HCs are high according to the analyzer, because HC is the result of unburned gasoline.

Answer C is correct. Both Technicians are correct.

Answer D is incorrect. Both Technicians are correct.

TASK B.1

17. A test light is connected between the negative side of the ignition coil and ground and the engine is cranked. Technician A says a flickering test light could be caused by a defective ignition module. Technician B says a flickering test light could be caused by a defective pickup coil. Who is correct?

   A. A only

   B. B only

   C. Both A and B

   D. Neither A nor B

Answer A is incorrect. If the test light flickers on and off while cranking, then the ignition module is doing what it is supposed to do.

Answer B is incorrect. If the test light flickers on and off while cranking, then the pickup coil is doing what it is supposed to do.

Answer C is incorrect. Neither Technician is correct.

Answer D is correct. Neither Technician is correct. When diagnosing a no-start with no spark, a test light should be placed between the negative terminal of the ignition coil and ground. Crank the engine and watch the test light. If the ignition control module and its trigger device are working correctly, then the test light will flicker on and off.

18. A multi-port fuel injection vehicle has a tip-in hesitation when warm. All of these could be the cause EXCEPT:

   A. Throttle position sensor (TPS)

   B. Canister purge control solenoid

   C. EGR

   D. Defective secondary air pump motor

Answer A is incorrect. If the TPS gets a bad spot on it, then the vehicle will hesitate.

Answer B is incorrect. If the purge control solenoid allows the canister to purge at idle, then a hesitation can occur on acceleration and tip-in.

Answer C is incorrect. A defective EGR valve can open to soon, causing an acceleration tip-in complaint.

Answer D is correct. A defective secondary air pump motor would cause excessive emissions but would not cause a tip-in complaint.

19. Technician A says that one of the first steps in diagnosing any EGR valve related concern is to check the vacuum and electrical connections. Technician B says that in many systems, the PCM uses other sensor inputs that could cause an EGR problem, and therefore DTCs should be corrected before replacing any EGR components. Who is correct?

   A. A only

   B. B only

   C. Both A and B

   D. Neither A nor B

Answer A is incorrect. Technician B is also correct.

Answer B is incorrect. Technician A is also correct.

Answer C is correct. Both Technicians are correct. A good visual inspection is part of any diagnostic routine and should be performed after the complaint is verified. The PCM uses many sensors to determine the amount of EGR flow to allow; thus, diagnostic trouble codes should be corrected before replacing any EGR components.

Answer D is incorrect. Both Technicians are correct.

OXYGEN SENSOR VOLTAGE VARIATIONS

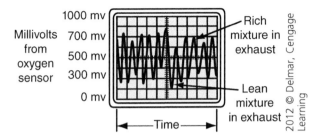

TASKS E.3,
E.4

20. Based on the O_2 (oxygen) sensor wave form shown in the illustration, which of the following is true?

   A. This represents a lean-biased condition.

   B. The O_2 sensor is functioning correctly.

   C. This represents a lazy oxygen sensor.

   D. This represents a rich-biased condition.

   Answer A is incorrect. A lean bias sensor would stay below 450 millivolts.

   Answer B is correct. The O_2 sensor is switching back and forth from about 800 to 100 millivolts.

   Answer C is incorrect. A lazy O_2 sensor would be switching, but at a much slower rate.

   Answer D is incorrect. A rich bias sensor would stay above 450 millivolts.

TASK A.2

21. Technician A says secondary air-injection systems must be monitored for proper operation on a vehicle certified as OBD II-compliant if equipped. Technician B says secondary air-injection systems are not monitored on OBD I-compliant vehicles. Who is correct?

   A. A only

   B. B only

   C. Both A and B

   D. Neither A nor B

   Answer A is incorrect. Technician B is also correct.

   Answer B is incorrect. Technician A is also correct.

   Answer C is correct. Both Technicians are correct. If an OBD II vehicle is equipped with a secondary air-injection system, then the PCM must run monitors on it to verify correct operation. Many OBD I vehicles were equipped with secondary air; however, the PCM could not monitor the system.

   Answer D is incorrect. Both Technicians are correct.

TASK B.6

22. A Hall-effect sensor is being tested. Technician A says the Hall-effect sensor should have a resistance value of 500 to 1,500 ohms. Technician B says a brass feeler gauge should be used to adjust a Hall-effect sensor. Who is correct?

   A. A only

   B. B only

   C. Both A and B

   D. Neither A nor B

   Answer A is incorrect. While the resistance value shown would be good for a pickup coil, a Hall-effect sensor cannot be checked for resistance.

   Answer B is incorrect. A brass feeler gauge is used to adjust a pickup coil, not a Hall-effect sensor.

   Answer C is incorrect. Neither Technician is correct.

   Answer D is correct. Neither Technician is correct. The best method to check a Hall-effect sensor for proper operation is to use a lab scope and view the wave form produced.

23. All of the following are possible causes of turbocharger failure EXCEPT:

    A. Poor intake air filtration
    B. Poor engine oil maintenance
    C. Poor engine cooling system maintenance
    D. Poor exhaust system maintenance.

TASK C.14

    Answer A is incorrect. The air filter must be checked and replaced as needed for maximum protection from airborne abrasion wear.

    Answer B is incorrect. While the engine oil lubricates and cools the engine, it also lubricates and cools the turbocharger.

    Answer C is incorrect. Turbochargers produce high temperature, so the cooling system is used to help transfer the heat away from the turbocharger.

    Answer D is correct. While poor exhaust system maintenance can lead to exhaust leak, it does not cause premature turbo failure.

24. Technician A says the EVAP system assists in the reduction of oxides of nitrogen (NOx). Technician B says the EVAP system prevents fuel vapors from escaping into the atmosphere. Who is correct?

    A. A only
    B. B only
    C. Both A and B
    D. Neither A nor B

TASK D.4.1

    Answer A is incorrect. The EGR system is in charge of reducing NOx, not the EVAP system.

    Answer B is correct. Only Technician B is correct. The EVAP system is used to prevent the escape of HC vapors into the atmosphere.

    Answer C is incorrect. Only Technician B is correct.

    Answer D is incorrect. Technician B is correct.

25. When testing for voltage drop in the power and ground distribution circuits, Technician A says the circuit being tested must be operating. Technician B says any corrosion adds unwanted resistance. Who is correct?

    A. A only
    B. B only
    C. Both A and B
    D. Neither A nor B

TASK E.5

    Answer A is incorrect. Technician B is also correct.

    Answer B is incorrect. Technician A is also correct.

    Answer C is correct. Both Technicians are correct. Any time a voltage drop test is being performed, the circuit being tested must be powered up and operating. Any resistance in the circuit will show up as a voltage drop.

    Answer D is incorrect. Both Technicians are correct.

TASK A.20

26. When the alternator belt and belt tension are satisfactory and the alternator output is low, Technician A says the alternator may be defective. Technician B says the problem could be high resistance in the alternator field circuit. Who is correct?

   A. A only

   B. B only

   C. Both A and B

   D. Neither A nor B

   Answer A is incorrect. Technician B is also correct.

   Answer B is incorrect. Technician A is also correct.

   Answer C is correct. Both Technicians are correct. A defective alternator and high resistance in the field circuit could cause low output from the alternator.

   Answer D is incorrect. Both Technicians are correct.

TASK B.3

27. A pickup coil resistance is being tested with an ohmmeter. Technician A says when the pickup coil leads are moved, an erratic ohmmeter reading indicates the need for replacement. Technician B says that an infinite ohmmeter reading between the pickup coil terminals is normal on some pickup coils. Who is correct?

   A. A only

   B. B only

   C. Both A and B

   D. Neither A nor B

   Answer A is correct. Only Technician A is correct. Any erratic ohmmeter readings when the pickup coil leads are moved indicates a partially open circuit or that it is intermittently making good connection and the pickup coil should be replaced.

   Answer B is incorrect. Pickup coils are magnetic pulse generating components and they all have a resistance value, usually between 500 to 1,500 ohms. An infinite reading on a pickup coil would indicate the need for replacement.

   Answer C is incorrect. Only Technician A is correct.

   Answer D is incorrect. Technician A is correct.

TASK C.4

28. While testing fuel pressure on a multi-port fuel injection engine, Technician A says there will always be a Schrader test port for fuel system testing. Technician B says that a high fuel pressure reading could be the result of a plugged return line. Who is correct?

   A. A only

   B. B only

   C. Both A and B

   D. Neither A nor B

   Answer A is incorrect. Not all vehicles have a Schrader valve for testing the fuel pressure. Some vehicles require the installation of a test port so a pressure gauge can be connected.

   Answer B is correct. Only Technician B is correct. If high fuel pressure is found, then a restricted or plugged fuel return line could be the cause.

   Answer C is incorrect. Only Technician B is correct.

   Answer D is incorrect. Technician B is correct.

29. A stuck air-switching valve that constantly sends air pump output to the exhaust manifold will most likely result in which of the following conditions?

    A. Poor fuel economy

    B. A constant rich oxygen sensor signal

    C. Engine spark knock

    D. The exhaust to overheat

    TASK D.3.1

    Answer A is correct. If the secondary air is constantly injected upstream into the exhaust manifold, then the extra oxygen tricks the O_2 sensor into sending a lean signal to the PCM, causing the PCM to command an increase in fuel.

    Answer B is incorrect. A constant lean oxygen sensor signal would be the result.

    Answer C is incorrect. This would have no effect on any spark knock.

    Answer D is incorrect. This would not cause the exhaust to overheat.

30. Technician A says some PCMs require a program chip to be transferred from the old PCM to the new PCM when replacing. Technician B says some PCMs require flash programming when replacing. Who is correct?

    A. A only

    B. B only

    C. Both A and B

    D. Neither A nor B

    TASK E.7

    Answer A is incorrect. Technician B is also correct.

    Answer B is incorrect. Technician A is also correct.

    Answer C is correct. Both Technicians are correct. Some PCMs require the programmable read only memory (PROM) chip be transferred from the old PCM when replacing. Some PCMs use an electrically erasable programmable read-only memory (EEPROM) chip that requires the PCM be flash programmed when replaced.

    Answer D is incorrect. Both Technicians are correct.

31. A vehicle emits a belt squeal when starting and on acceleration. Technician A says the alternator bearings may be defective. Technician B says the alternator belt automatic tensioner may be defective. Who is correct?

    A. A only

    B. B only

    C. Both A and B

    D. Neither A nor B

    TASK A.22

    Answer A is incorrect. A defective bearing would make noise all the time, not just at start up or acceleration.

    Answer B is correct. Only Technician B is correct. If the tensioner fails to keep the belt tight, then the belt could squeal on start up or acceleration.

    Answer C is incorrect. Only Technician B is correct.

    Answer D is incorrect. Technician B is correct.

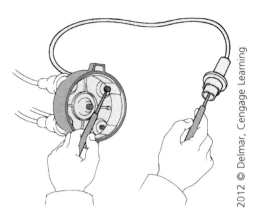

TASK B.4

32. Refer to the above illustration. Technician A says secondary circuit resistance is being checked. Technician B says high resistance in the shown component could lead to ignition coil failure. Who is correct?

A. A only
B. B only
C. Both A and B
D. Neither A nor B

Answer A is incorrect. Technician B is also correct.

Answer B is incorrect. Technician A is also correct.

Answer C is correct. Both Technicians are correct. Secondary circuit resistance is being checked. High secondary circuit resistance can lead to ignition coil and ignition module failure.

Answer D is incorrect. Both Technicians are correct.

TASK C.10

33. When a vacuum leak is suspected for a high idle complaint, the preferred testing method would be:

A. Propane
B. Smoke
C. Water
D. Throttle body cleaner

Answer A is incorrect. Propane can be used to locate a vacuum leak, but it is not the preferred method.

Answer B is correct. The best way and preferred method to locate a vacuum leak is with a smoke machine.

Answer C is incorrect. Water can damage components and allow the formation of corrosion.

Answer D is incorrect. Throttle body cleaner can be a fire hazard around a running engine.

34. Technician A says the PCM controls the amount of canister purge based on other inputs. Technician B says the EVAP system controls the amount of CO emissions produced by the engine. Who is correct?

TASK D.4.1

    A. A only
    B. B only
    C. Both A and B
    D. Neither A nor B

    Answer A is correct. Only Technician A is correct. The PCM controls the canister purge solenoid and allows canister purging based on inputs from other sensors like the ECT, TPS and heated O_2.

    Answer B is incorrect. The evaporative emissions system controls the escape of HC into the atmosphere, not CO.

    Answer C is incorrect. Only Technician A is correct.

    Answer D is incorrect. Technician A is correct.

35. A TPS is being tested with a digital multimeter (DMM). Technician A says the voltage on the signal wire should be around 1.0 volt or less at idle. Technician B says the voltage reading on the reference wire should be around 5 volts. Who is correct?

TASK E.4

    A. A only
    B. B only
    C. Both A and B
    D. Neither A nor B

    Answer A is incorrect. Technician B is also correct.

    Answer B is incorrect. Technician A is also correct.

    Answer C is correct. Both Technicians are correct. The TPS is a 5-volt sensor. Three wires are used on most TPSs include a 5-volt reference, a signal, and a ground. If the signal wire is back-probed with the key on, then the voltage reading should be around 1.0 volt or less at idle and around 4.5 volts at wide-open throttle.

    Answer D is incorrect. Both Technicians are correct.

36. Which of the following conditions is LEAST LIKELY to be diagnosed with an emissions analyzer?

TASK A.10

    A. Cylinder misfire
    B. Cylinder efficiency
    C. Head gasket integrity
    D. Exhaust system leakage

    Answer A is incorrect. A cylinder misfire can be diagnosed using an emissions analyzer. A misfiring cylinder will cause excessive HC gases.

    Answer B is incorrect. Cylinder efficiency can be determined by looking at the carbon dioxide (CO_2) readings. The higher, the better.

    Answer C is incorrect. The radiator can be sniffed with the analyzer exhaust probe. A reading of hydrocarbon gas in the radiator indicates a leaking head gasket.

    Answer D is correct. While an exhaust leak can cause the readings to be incorrect, an emissions analyzer is not used to detect exhaust leaks.

TASK C.8

37. Technician A says an injector with a lower pressure drop during an injector pressure-drop test indicates a rich-running injector. Technician B says an injector with a lower pressure drop could be leaking. Who is correct?

　　A.　A only

　　B.　B only

　　C.　Both A and B

　　D.　Neither A nor B

Answer A is incorrect. A lower pressure drop indicates a restriction or clogged injector.

Answer B is incorrect. A leak would be indicated by a higher pressure drop, not a lower one.

Answer C is incorrect. Neither Technician is correct.

Answer D is correct. Neither Technician is correct. When testing injectors for pressure drop, each injector is pulsed the same exact amount, and the amount of pressure drop is noted. If an injector does not drop the same amount as the others, then that injector has a restriction.

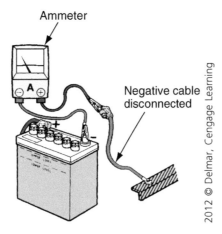

TASK A.17

38. A DMM set in the milliamp position is connected in series between the negative battery terminal and the negative cable, as shown. What is being measured?

　　A.　Voltage drop

　　B.　Open circuit voltage

　　C.　Parasitic drain

　　D.　Charging system amperage

Answer A is incorrect. If voltage drop was being measured, then the meter would be set to voltage and connected in parallel.

Answer B is incorrect. Open circuit voltage is measured with the meter on voltage and connected parallel.

Answer C is correct. *Parasitic drain* measures the amount of current being used with the key off.

Answer D is incorrect. Charging system amperage in measured with the meter set on amps, not milliamps and an amp clamp.

39. When replacing a PROM, Technician A says that you should always ground yourself to the vehicle. Technician B says that the PCM connector should be disconnected with the key on to prevent static charge. Who is correct?

TASK E.7

    A. A only

    B. B only

    C. Both A and B

    D. Neither A nor B

    Answer A is correct. Only Technician A is correct. You should always ground yourself to the vehicle when replacing the PROM. A special wrist ground strap can be used for this.

    Answer B is incorrect. The PCM connector should never be disconnected with the key on.

    Answer C is incorrect. Only Technician A is correct.

    Answer D is incorrect. Technician A is correct.

40. Technician A says the throttle body must be completely disassembled before soaking it in cleaning solvent. Technician B says some throttle bodies cannot be cleaned in solvent. Who is correct?

TASK C.7

    A. A only

    B. B only

    C. Both A and B

    D. Neither A nor B

    Answer A is incorrect. Technician B is also correct.

    Answer B is incorrect. Technician A is also correct.

    Answer C is correct. Both Technicians are correct. It a throttle body service is to be performed, the throttle body should be disassembled, removing all electrical components. Not all throttle bodies can be serviced; some manufacturers require the replacement of the throttle body if it gets carboned up.

    Answer D is incorrect. Both Technicians are correct.

41. Technician A says that an overfilled crankcase can cause hydraulic lifter noise due to oil aeration. Technician B says using motor oil with a viscosity rating that is too low can cause hydraulic lifter noise. Who is correct?

TASK A.3

    A. A only

    B. B only

    C. Both A and B

    D. Neither A nor B

    Answer A is incorrect. Technician B is also correct.

    Answer B is incorrect. Technician A is also correct.

    Answer C is correct. Both Technicians are correct. Aeration is the formation of air bubbles. Overfilling the crankcase with oil can cause this aeration, and the air bubbles can lead to lifter noise. Another cause of lifter noise is the use of engine oil that is too thin or the viscosity too low.

    Answer D is incorrect. Both Technicians are correct.

TASK E.4

42. Technician A says that a DMM can be used to check an oxygen sensor. Technician B says that to check an oxygen sensor, you can use an oscilloscope. Who is correct?

    A. A only
    B. B only
    C. Both A and B
    D. Neither A nor B

    Answer A is incorrect. Technician B is also correct.

    Answer B is incorrect. Technician A is also correct.

    Answer C is correct. Both Technicians are correct. A good oxygen sensor should rapidly switch from 100 to 900 millivolts. This can be tested with a DMM set on the 2-volt scale or with oscilloscope.

    Answer D is incorrect. Both Technicians are correct.

TASK E.4

43. When testing the engine coolant temperature sensor circuit, all of the following are used EXCEPT:

    A. Scan tool data
    B. Resistance value
    C. Voltage value
    D. Amperage value

    Answer A is incorrect. Scan tool data is valuable information when testing the engine coolant temperature sensor circuit.

    Answer B is incorrect. Resistance value is valuable information when testing the engine coolant temperature sensor circuit.

    Answer C is incorrect. Voltage value is valuable information when testing the engine coolant temperature sensor circuit.

    Answer D is correct. The amperage value is not used when testing the engine coolant temperature sensor circuit.

TASK A.15

44. A vehicle is being diagnosed for an overheating complaint while at idle. Technician A says the electric cooling fan may be defective. Technician B says that the thermostat may be opening too soon. Who is correct?

    A. A only
    B. B only
    C. Both A and B
    D. Neither A nor B

    Answer A is correct. Only Technician A is correct. The fan is used to circulate air through the radiator at a stop or when idling and temperatures are high. A defective fan could cause the overheating complaint.

    Answer B is incorrect. If the thermostat was opening too soon, then the vehicle would take a long time to reach operating temperature, if it reached it at all.

    Answer C is incorrect. Only Technician A is correct.

    Answer D is incorrect. Technician A is correct.

45. Resistance on a negative-temperature coefficient coolant sensor is being tested against specifications. Technician A says a resistance reading lower than specifications would send a signal indicating a warmer-than-actual engine temperature. Technician B says if the resistance is lower than specifications, the engine may exhibit hard starting when warm. Who is correct?

TASK E.4

A. A only

B. B only

C. Both A and B

D. Neither A nor B

Answer A is correct. Only Technician A is correct. With a negative-temperature coefficient sensor, as temperature goes up, resistance goes down. As the engine warms up, the resistance of the ECT goes down. A lower-than-specification reading would send a signal indicating a warmer-than-actual engine temperature.

Answer B is incorrect. On a warm engine, this would have little if any effect on starting. On a cold engine, however, this could cause hard starting.

Answer C is incorrect. Only Technician A is correct.

Answer D is incorrect. Technician A is correct.

46. A vacuum gauge indicates low vacuum (12 inches of mercury). Technician A says late valve timing will cause a low vacuum reading. Technician B says to connect the gauge to a venturi vacuum port. Who is correct?

TASK A.5

A. A only

B. B only

C. Both A and B

D. Neither A nor B

Answer A is correct. Only Technician A is correct. Late valve timing will cause the valves to open and close at the wrong time in relation to piston position. This will cause a low vacuum reading.

Answer B is incorrect. The vacuum gauge should be connected to manifold vacuum, not venturi vacuum.

Answer C is incorrect. Only Technician A is correct.

Answer D is incorrect. Technician A is correct.

47. Technician A says the PCM will increase the fuel injector pulse width if there is no oxygen in the exhaust. Technician B says if there is a lean condition, then the short-term fuel trim (SFT) will show a minus value on the scan tool. Who is correct?

TASK E.3

A. A only

B. B only

C. Both A and B

D. Neither A nor B

Answer A is incorrect. If there is no oxygen in the exhaust, then this indicates a rich condition. The PCM will decrease fuel injector pulse width, not increase it.

Answer B is incorrect. If there is a lean condition, then the SFT will show a plus sign indicating it is adding fuel.

Answer C is incorrect. Neither Technician is correct.

Answer D is correct. Neither Technician is correct. The PCM increases or decreases the fuel injector pulse width to control the air/fuel mixture based on what the oxygen sensor signal sends.

TASK A.21

48. Which of the follow is the LEAST LIKELY cause of alternator drive belt noise?

    A. Defective belt tensioner
    B. Glazed belt
    C. Belt alignment
    D. Over-tighten belt

    Answer A is incorrect. If the belt tensioner fails to keep the belt at the proper tension, then the belt can become noisy.

    Answer B is incorrect. If a belt becomes glazed, then it will slip and cause noise.

    Answer C is incorrect. Belt alignment problems will cause premature belt failure as well as noise.

    Answer D is correct. An overtightened belt can cause premature component failure, such as water pump and alternator, but does not cause noise.

TASK E.1

49. Which of the follow diagnostic trouble codes are Type A codes and should be troubleshot first?

    A. Transmission
    B. Vehicle speed control
    C. Misfire control
    D. Non-emissions-related

    Answer A is incorrect. A transmission code is a Type C or D code.

    Answer B is incorrect. A vehicle speed control code is a Type C or D code.

    Answer C is correct. A misfire code is a Type A code and will illuminate the MIL (malfunction indicator lamp) on the first trip failure.

    Answer D is incorrect. Non-emissions-related codes are Type C or D codes.

TASK E.5

50. What is the maximum voltage drop for a PCM ground circuit?

    A. 0.10 volts
    B. 0.30 volts
    C. 0.20 volts
    D. 1.0 volts

    Answer A is correct. A computer ground should have no more than 0.10 voltage drop.

    Answer B is incorrect. 0.30 volts is too much for a computer circuit; 0.30 is the max for a switch.

    Answer C is incorrect. 0.20 volts is the max for a wire or cable.

    Answer D is incorrect. 1.0 volts is too much for any circuit.

PREPARATION EXAM 4—ANSWER KEY

1.	A	21.	D	41.	C
2.	C	22.	B	42.	C
3.	B	23.	D	43.	C
4.	B	24.	C	44.	A
5.	C	25.	C	45.	B
6.	B	26.	C	46.	C
7.	A	27.	B	47.	B
8.	C	28.	C	48.	A
9.	C	29.	C	49.	C
10.	C	30.	B	50.	C
11.	C	31.	C		
12.	A	32.	C		
13.	A	33.	B		
14.	D	34.	D		
15.	C	35.	C		
16.	A	36.	C		
17.	C	37.	A		
18.	C	38.	C		
19.	C	39.	D		
20.	B	40.	B		

PREPARATION EXAM 4—EXPLANATIONS

1. Technician A says if the PCV valve is stuck closed, then excessive crankcase pressure forces blowy gases through the clean air hose into the air filter. Technician B says if the positive crankcase ventilation (PCV) valve is stuck open, then excessive airflow through the valve causes a rich air/fuel ratio. Who is correct?

TASK D.1.1

A. A only

B. B only

C. Both A and B

D. Neither A nor B

Answer A is correct. Only Technician A is correct. If the PCV valve were to stick closed, then excessive blow-by gases would build in the engine, forcing those gases (and often oil) through the air cleaner hose and into the air filter.

Answer B is incorrect. A stuck-open PCV valve will cause the air/fuel mixture to lean out, not richen up.

Answer C is incorrect. Only Technician A is correct.

Answer D is incorrect. Technician A is correct.

TASK B.3

2. Technician A says that a 12-volt test light connected between the negative side of an ignition coil and ground that blinks on and off during cranking confirms the primary circuit is being switched. Technician B says any voltage drops greater than .3 volts in the primary circuit can reduce secondary circuit kV output. Who is correct?

   A. A only
   B. B only
   C. Both A and B
   D. Neither A nor B

   Answer A is incorrect. Technician B is also correct.

   Answer B is incorrect. Technician A is also correct.

   Answer C is correct. Both Technicians are correct. If the test light blinks when the engine is cranked, it verifies the primary circuit is being switched to ground and then off. Any voltage drop in the primary circuit will affect the amount of secondary voltage produced. A 1-volt voltage drop in the primary can lower the secondary voltage by as much as 10,000 volts.

   Answer D is incorrect. Both Technicians are correct.

TASK A.3

3. A 6-cylinder engine is making a loud metallic knocking that gets louder as the engine warms up or if the throttle is quickly snapped open. The noise almost disappears when the spark for cylinder #2 is shorted to ground. Technician A says the problem could be a cracked flex plate. Technician B says the problem is most likely a loose connecting rod bearing. Who is correct?

   A. A only
   B. B only
   C. Both A and B
   D. Neither A nor B

   Answer A is incorrect. A cracked flex plate making noise would not be affected by shorting cylinder #2.

   Answer B is correct. Only Technician B is correct. If a knock gets quieter when the spark is shorted for the cylinder with a loose rod bearing, then the rod bearings are most likely the cause.

   Answer C is incorrect. Only Technician B is correct.

   Answer D is incorrect. Technician B is correct.

TASK C.1

4. All of the following symptoms can be caused by low fuel pressure EXCEPT:

   A. Engine surge
   B. Strong sulfur smell from the exhaust
   C. Low engine power
   D. Limited top speed

   Answer A is incorrect. An engine surge is one of the symptoms of low fuel pressure.

   Answer B is correct. A strong sulfur smell from the exhaust is a symptom of a rich mixture. Low fuel pressure would cause a lean mixture.

   Answer C is incorrect. A vehicle with low fuel pressure will have low engine power.

   Answer D is incorrect. A vehicle with low fuel pressure will have limited top vehicle speeds.

5. All of the following statements about powertrain control module (PCM) inputs are true EXCEPT:

TASK E.5

   A. High-impedance digital volt/ohm meters may be used for diagnosis.

   B. The O_2 sensor produces very low voltage.

   C. An analog meter may be used for diagnosis.

   D. Most inputs use a 5-volt reference voltage.

   Answer A is incorrect. A digital volt/ohm meter is the preferred meter to use when diagnosing computer circuits.

   Answer B is incorrect. Oxygen (O_2) sensors produce a voltage signal of 0 to 1 volts.

   Answer C is correct. Most analog meters are low impedance, meaning they have low internal resistance. These types of meters can draw more amperage than a circuit can handle, so they are not used when testing computer circuits.

   Answer D is incorrect. Computers work off of 5 volts, so most input sensors use a 5-volt reference signal.

6. A technician is performing a compression test. Which statement below is most likely true?

TASK A.7

   A. A low reading on one cylinder may be caused by a vacuum leak at that cylinder.

   B. All cylinders reading even, but lower than normal, may be caused by a slipped timing chain.

   C. Low readings on two adjacent cylinders may be caused by carbon buildup.

   D. All cylinders with higher than normal readings could be caused by a blown head gasket.

   Answer A is incorrect. A vacuum leak would not affect a compression test.

   Answer B is correct. A slipped timing chain will usually show low and even compression on all the cylinders.

   Answer C is incorrect. Low readings on two adjacent cylinders indicates a cracked head or blown head gasket.

   Answer D is incorrect. If a cylinder shows a compression reading higher than normal, then carbon buildup is usually the cause.

TASK A.15

7. Technician A says the thermostatic coil controls the opening and closing of the orifice inside the coupling on a viscous fan clutch. Technician B says when the thermostatic coil is cold, the orifice is open on a viscous fan clutch. Who is correct?

    A. A only
    B. B only
    C. Both A and B
    D. Neither A nor B

    Answer A is correct. Only Technician A is correct. The viscous fan clutch is designed to give maximum airflow when hot and minimum air flow when by opening and closing a orifice inside the viscous clutch.

    Answer B is incorrect. The orifice is opened when the thermostatic viscous fan clutch is hot and closed when cold.

    Answer C is incorrect. Only Technician A is correct.

    Answer D is incorrect. Technician A is correct.

TASK B.1

8. Technician A says when testing an ignition related no-start problem, the technician should always check for available spark at an ignition wire first. Technician B says if a test light is connected to the negative side of the coil while cranking and the test light flickers, then the secondary ignition system needs to be tested. Who is correct?

    A. A only
    B. B only
    C. Both A and B
    D. Neither A nor B

    Answer A is incorrect. Technician B is also correct.

    Answer B is incorrect. Technician A is also correct.

    Answer C is correct. Both Technicians are correct. One of the first steps in diagnosing a no-start is to check for spark at one of the secondary ignition wires. With a test light connected between the negative terminal of the ignition coil and ground, crank the engine and watch for a flashing test light. This indicates the primary ignition system is switching correctly, so the secondary should be checked.

    Answer D is incorrect. Both Technicians are correct.

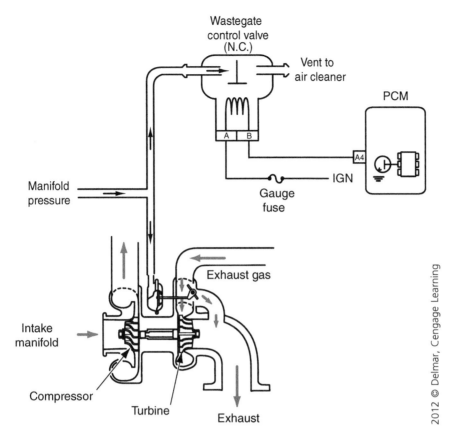

9. Refer to the above illustration. All of the following conditions could reduce engine power during a boost condition EXCEPT:

A. Damaged compressor fins

B. Wastegate valve stuck open

C. A blown gauge fuse

D. A leak between the turbocharger and exhaust manifold

TASK C.14

Answer A is incorrect. If the compressor has damaged fins, then the amount of boost pressure would be limited.

Answer B is incorrect. If the wastegate is stuck open, then the boost pressure will be lowered.

Answer C is correct. A blown gauge fuse would prevent the wastegate control from opening. This would cause an overboost, because this valve is normally closed.

Answer D is incorrect. A leak between the turbocharger and exhaust manifold would reduce turbocharger output.

TASK D.2.3

10. Technician A says that when vacuum is applied to some exhaust gas recirculation (EGR) valves with the engine idling, the EGR valve should open and idle should become erratic. Technician B says that some EGR valves can be opened with a scan tool. Who is correct?

   A. A only
   B. B only
   C. Both A and B
   D. Neither A nor B

   Answer A is incorrect. Technician B is also correct.

   Answer B is incorrect. Technician A is also correct.

   Answer C is correct. Both Technicians are correct. Most diaphragm-type EGR valves can be tested for proper movement by applying vacuum to the valve at idle and watching for a drop in engine idle. A scan tool also can be used to open the EGR valve for testing purposes if the vehicle is equipped with electronic EGR valves.

   Answer D is incorrect. Both Technicians are correct.

TASK E.4

11. Technician A says to use a scan tool to verify throttle position sensor input and related DTCs (diagnostic trouble codes). Technician B says coolant-temperature sensor input is used to help determine open- and closed-loop status. Who is correct?

   A. A only
   B. B only
   C. Both A and B
   D. Neither A nor B

   Answer A is incorrect. Technician B is also correct.

   Answer B is incorrect. Technician A is also correct.

   Answer C is correct. Both Technicians are correct. A scan tool can be used to monitor TPS (throttle position sensor) readings in voltage as well as degree of opening. If an open or short occurs in the TPS circuit, then a DTC will be stored in the PCM. The coolant-temperature sensor is one of the inputs that determine when the engine enters closed-loop status.

   Answer D is incorrect. Both Technicians are correct.

TASK E.4

12. Technician A says a scan tool can be used to check for O_2 sensor codes and operation. Technician B says. The O_2 sensor can be removed and tested under for proper operation. Who is correct?

   A. A only
   B. B only
   C. Both A and B
   D. Neither A nor B

   Answer A is correct. Only Technician A is correct. A scan tool can be used to monitor the operation of the O_2 sensors on a vehicle. The voltage is monitored and should be switching back and forth between 100 millivolts and 900 millivolts.

   Answer B is incorrect. The oxygen sensor cannot be accurately diagnosed off the vehicle.

   Answer C is incorrect. Only Technician A is correct.

   Answer D is incorrect. Technician A is correct.

13. Technician A says that an oscilloscope can be used to watch the mass air flow (MAF) sensor signal switch from idle to wide-open throttle (WOT) status. Technician B says fuel injectors can only be tested using an ohmmeter. Who is correct?

TASK E.8,
C.8

   A. A only

   B. B only

   C. Both A and B

   D. Neither A nor B

   Answer A is correct. Only Technician A is correct. An oscilloscope is used to test the operation on the MAF. Depending on what type of MAF sensor is used, it will produce a digital signal or a frequency.

   Answer B is incorrect. Checking the resistance of an injector is one test that is performed, but not the only test. A pressure drop test is another test that can be performed.

   Answer C is incorrect. Only Technician A is correct.

   Answer D is incorrect. Technician A is correct.

14. While performing a valve adjustment, Technician A says the crankshaft must be placed in position so the piston is at TDC (top dead center) on the exhaust stroke. Technician B says that adjusting valves with too much clearance may cause rough running and burnt valves. Who is correct?

TASK A.11

   A. A only

   B. B only

   C. Both A and B

   D. Neither A nor B

   Answer A is incorrect. The piston should be at TDC of the compression stroke; this ensures the valves are fully closed.

   Answer B is incorrect. If the valves were adjusted too loose or with too much clearance, then they would be noisy.

   Answer C is incorrect. Neither Technician is correct.

   Answer D is correct. Neither Technician is correct. When adjusting valve, some manufacturers specify on a cold engine and some on a hot engine. The piston must be at TDC on the compression stroke.

15. Technician A says an ignition coil should be tested for both primary and secondary winding resistance. Technician B says available coil output can be tested with an oscilloscope. Who is correct?

TASK B.5

   A. A only

   B. B only

   C. Both A and B

   D. Neither A nor B

   Answer A is incorrect. Technician B is also correct.

   Answer B is incorrect. Technician A is also correct.

   Answer C is correct. Both Technicians are correct. When testing an ignition coil, the primary and secondary resistance should be measured. The available voltage or maximum voltage can be checked using an oscilloscope.

   Answer D is incorrect. Both Technicians are correct.

TASK C.8

16. Technician A says if an engine uses low-resistance injectors with less than 3 ohms resistance, then the PCM will use a current-limiting or peak and hold injector driver to operate the fuel injector. Technician B says using a current-limiting or peak and hold injector increases fuel injector noise. Who is correct?

A. A only

B. B only

C. Both A and B

D. Neither A nor B

Answer A is correct. Only Technician A is correct. An injector with less than 3 ohms resistance will be controlled with a current-limiting or peak and hold injector driver.

Answer B is incorrect. A peak and hold injector does not produce more noise than any other type of injector.

Answer C is incorrect. Only Technician A is correct.

Answer D is incorrect. Technician A is correct.

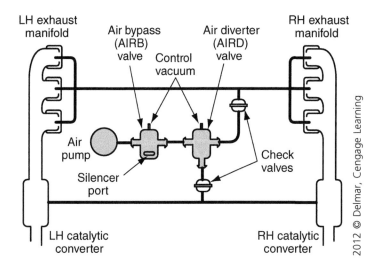

17. Refer to the above illustration. Technician A says the check valves allow air in the exhaust without exhaust getting in the air diverter valve. Technician B says the air bypass valve prevents secondary air from entering the exhaust on deceleration. Who is correct?

TASK D.3.3

A. A only

B. B only

C. Both A and B

D. Neither A nor B

Answer A is incorrect. Technician B is also correct.

Answer B is incorrect. Technician A is also correct.

Answer C is correct. Both Technicians are correct. On a cold engine, the secondary air is diverted upstream in the exhaust manifold. Once the engine enters closed loop, the secondary air is diverted downstream to the catalytic converters. On deceleration, the secondary air is bypassed to the atmosphere through the bypass valve. The check valves prevent the exhaust from entering the diverter valve, which would be damaged by the hot exhaust.

Answer D is incorrect. Both Technicians are correct.

18. Technician A says a manifold absolute pressure (MAP) sensor should be able to hold vacuum during a test. Technician B says some MAP sensors produce an analog voltage signal while others produce a digital square wave signal. Who is correct?

TASK E.4

A. A only

B. B only

C. Both A and B

D. Neither A nor B

Answer A is incorrect. Technician B is also correct.

Answer B is incorrect. Technician A is also correct.

Answer C is correct. Both Technicians are correct. MAP sensors can be analog or digital. Analog typically produces a direct-current voltage signal while digital sensors produce a frequency. Either type should be able to hold a vacuum without leak down.

Answer D is incorrect. Both Technicians are correct.

19. The result of a battery load test done at 78°F with a carbon pile load tester is 8.9 volts. Technician A says minimum load test voltage is 9.6 volts, and this result is unacceptable. Technician B says that you should compare the results to the tables from the tool or battery manufacturer for temperature correction. Who is correct?

TASK A.17

A. A only

B. B only

C. Both A and B

D. Neither A nor B

Answer A is incorrect. Technician B is also correct.

Answer B is incorrect. Technician A is also correct.

Answer C is correct. Both Technicians are correct. When load testing the battery, half of the battery's rated cold cranking amps is applied to the battery for 15 seconds. For temperatures above 70°F, the voltage should not fall below 9.6 volts. If the battery is being tested at temperatures below 70°F, then a temperature correction chart must be used.

Answer D is incorrect. Both Technicians are correct.

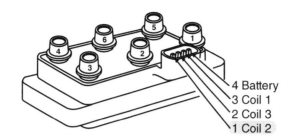

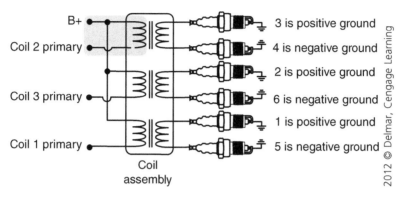

TASK B.5

20. Technician A says the schematic in the above illustration is for a COP (coil-on plug) ignition system. Technician B says if coil primary #2 became open, then two cylinders would be killed. Who is correct?

A. A only
B. B only
C. Both A and B
D. Neither A nor B

Answer A is incorrect. COPs have individual coils for each cylinder. The schematic shows two cylinders per coil.

Answer B is correct. Only Technician B is correct. The schematic is for a waste spark ignition system. Two cylinders are fired from one coil. If the primary circuit becomes open, then it will affect two cylinders.

Answer C is incorrect. Only Technician B is correct.

Answer D is incorrect. Technician B is correct.

TASK C.1

21. Which of the following is the LEAST LIKELY cause of poor fuel mileage on a vehicle with SFI (sequential fuel injection)?

A. Restricted pressure regulator
B. A defective oxygen sensor
C. Restricted exhaust
D. Defective speed sensor

Answer A is incorrect. A restricted fuel pressure regulator would cause high fuel pressure and would cause a loss of fuel mileage.

Answer B is incorrect. A defective oxygen sensor would send a false signal or no signal at all causing the PCM to increase the pulse width causing poor fuel mileage.

Answer C is incorrect. A restricted exhaust would cause the engine to work harder than normal causing poor fuel mileage.

Answer D is correct. A defective speed sensor would affect the speed control and speedometer, but would not affect fuel mileage.

22. All of the following could cause an EVAP system large leak code to be set EXCEPT:

    A. Loose gas cap
    B. Leaking injector O-ring
    C. Cracked purge control solenoid
    D. Cracked canister vent hose

TASK D.4.2

Answer A is incorrect. A loose cap will cause a large EVAP leak code.

Answer B is correct. The injector is not part of the EVAP system. A leaking o-ring would cause a drivability problem, but would not set a large leak code.

Answer C is incorrect. If the purge solenoid was cracked, then hydrocarbons could escape through the crack, and an EVAP large leak would be set.

Answer D is incorrect. A crack anywhere in the vent hose to the canister would cause a large leak code to be set.

23. Technician A says that for non-emissions-related DTCs, you can replace the component without using the flow chart. Technician B says depending on the DTC set, some steps of the flow chart can be bypassed. Who is correct?

    A. A only
    B. B only
    C. Both A and B
    D. Neither A nor B

TASK E.2

Answer A is incorrect. It is never ok to just replace a component without diagnosing the failure and bypassing the flow chart.

Answer B is incorrect. No matter what the code is, the flow chart should be followed step by step to ensure a correct diagnosis.

Answer C is incorrect. Neither Technician is correct.

Answer D is correct. Neither Technician is correct. When troubleshooting diagnostic trouble codes, a diagnostic flow chart should be used to ensure proper diagnostics.

24. Low battery or system voltage can cause all of the following EXCEPT:

    A. A code to be set
    B. Increased idle RPM
    C. Increased steering effort
    D. Poor drivability at high speeds

TASK A.17

Answer A is incorrect. Low battery or system voltage will cause a diagnostic code to be set.

Answer B is incorrect. If battery voltage gets low, then the PCM will increase the idle RPM.

Answer C is correct. Low battery voltage will not cause the steering effort to increase.

Answer D is incorrect. Low battery voltage will cause high speed drivability problems because of low available secondary voltage.

25. All of the following could cause a cylinder misfire diagnostic code to be set EXCEPT:

    A. Low fuel pump pressure
    B. Secondary insulation breakdown
    C. Worn cam shaft lobe
    D. Erratic crank shaft sensor signal

    Answer A is incorrect. Low fuel pressure will cause a cylinder misfire code to be set.

    Answer B is incorrect. A breakdown of the secondary system insulation would allow arcing and would cause a misfire code to be set.

    Answer C is correct. The misfire monitor reads the crankshaft signal to determine a miss. A cylinder that is missing will cause the engine to slow down when that cylinder is supposed to fire. A worn cam lobe will not cause the engine to slow down.

    Answer D is incorrect. An erratic crankshaft sensor signal will cause the PCM to think a misfire is taking place and set a code.

26. Technician A says that nylon fuel line should be routed in a way to prevent kinking. Technician B says nylon fuel line can be repaired on some vehicles; if not, the entire line must be replaced. Who is correct?

    A. A only
    B. B only
    C. Both A and B
    D. Neither A nor B

    Answer A is incorrect. Technician B is also correct.

    Answer B is incorrect. Technician A is also correct.

    Answer C is correct. Both Technicians are correct. Most late-model vehicles use nylon fuel line. When replacing nylon fuel line, care must be taken not to cause a kink, which restricts fuel flow. Sometimes the nylon fuel line is repairable. Repair can only be made with a special nylon fuel line repair kit.

    Answer D is incorrect. Both Technicians are correct.

27. Technician A says that when vacuum is applied to the EGR valve with the engine idling, the EGR valve should open and idle should increase. Technician B says that a diagnosis of the EGR valve should be done with the engine idling. Who is correct?

    A. A only
    B. B only
    C. Both A and B
    D. Neither A nor B

    Answer A is incorrect. One test to perform on the EGR system is to apply a vacuum to the EGR valve at idle and watch the idle RPM. If the EGR valve is working, then the engine idle will become erratic or the engine may even die. The idle will not increase.

    Answer B is correct. Only Technician B is correct. The best engine speed to test an EGR system is at idle. Testing the EGR valve at a higher engine RPM might hide the problem.

    Answer C is incorrect. Only Technician B is correct.

    Answer D is incorrect. Technician B is correct.

28. Technician A says diagnostic trouble codes should be erased using a scan tool. Technician B says the monitor for the related fault should be run before the vehicle is returned to the customer. Who is correct?

TASK E.9

A. A only

B. B only

C. Both A and B

D. Neither A nor B

Answer A is incorrect. Technician B is also correct.

Answer B is incorrect. Technician A is also correct.

Answer C is correct. Both Technicians are correct. The best method for erasing diagnostic trouble codes is to use a scan tool. Before the vehicle is returned to the customer, however, the monitor for the fault that was repaired should be run to verify the repair.

Answer D is incorrect. Both Technicians are correct.

29. Technician A says to repeat the pressure test after repairs are made to the cooling system to ensure that all leaks are found. Technician B says a cooling system pressure test should include testing the radiator cap. Who is correct?

TASK A.14

A. A only

B. B only

C. Both A and B

D. Neither A nor B

Answer A is incorrect. Technician B is also correct.

Answer B is incorrect. Technician A is also correct.

Answer C is correct. Both Technicians are correct. When repairing a cooling system leak, the system should be rechecked after the repair is made to ensure that no other problems exist. The radiator cap should be included in the cooling system pressure test.

Answer D is incorrect. Both Technicians are correct.

30. The PCM sends a digital signal to the ignition control module to control which of the following?

TASK B.7

A. RPM

B. Ignition timing

C. Cylinder identification

D. EGR

Answer A is incorrect. The ignition control module does not control RPM.

Answer B is correct. After the engine starts, the PCM takes control of the timing by sending a digital signal to the ignition module.

Answer C is incorrect. The cam sensor is responsible for cylinder identification.

Answer D is incorrect. The ignition control module does not control the EGR system.

TASK C.4

31. Technician A says a fuel-pressure test is performed to test fuel pump operation. Technician B says it is possible to have a good fuel pressure reading and insufficient fuel flow. Who is correct?

A. A only

B. B only

C. Both A and B

D. Neither A nor B

Answer A is incorrect. Technician B is also correct.

Answer B is incorrect. Technician A is also correct.

Answer C is correct. Both Technicians are correct. A fuel pressure test is performed to test the ability of the fuel pump to produce the correct pressure needed to operate the vehicle. A fuel pump can have good pressure and still not have enough flow or volume. A restricted fuel filter is usually the cause of this problem.

Answer D is incorrect. Both Technicians are correct.

TASK D.4.3

32. Technician A says charcoal canister filters are no longer serviceable. Technician B says gas caps with pressure and vacuum valves must be checked for leakage with a pressure tester. Who is correct?

A. A only

B. B only

C. Both A and B

D. Neither A nor B

Answer A is incorrect. Technician B is also correct.

Answer B is incorrect. Technician A is also correct.

Answer C is correct. Both Technicians are correct. Many years ago, the charcoal canister had a fiber-type filter that was replaced during a tune-up, but new late-model vehicles no longer have a filter that is serviceable. Canisters today are sealed units. With the enhanced EVAP system, the need for gas caps that seal well is crucial. When diagnosing an EVAP leak code, a special pressure tester is used to test the sealing action of the gas cap.

Answer D is incorrect. Both Technicians are correct.

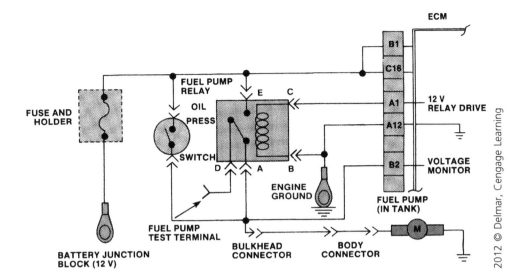

33. A vehicle with the fuel system shown in the above illustration is being diagnosed for a long cranking before start complaint. Technician A says the fuse could be blown, causing the long crank time before staring complaint. Technician B says this system type uses the oil pressure circuit as a back-up to the fuel pump relay circuit. Who is correct?

TASK C.5

A. A only

B. B only

C. Both A and B

D. Neither A nor B

Answer A is incorrect. If the fuse was blown, then the vehicle would not start at all, because the fuel pump would not get any voltage.

Answer B is correct. Only Technician B is correct. On this circuit, the oil pressure circuit is being used as a back up to the fuel pump control circuit. If the fuel pump relay fails, then the fuel pump will still energize once the oil pressure is high enough to close the oil pressure switch.

Answer C is incorrect. Only Technician B is correct.

Answer D is incorrect. Technician B is correct.

34. All of the following conditions could result in a voltage drop in a circuit EXCEPT:

A. Spread terminals in a connector

B. Broken strands in a stranded wire

C. Greenish corrosion builds up in a connector

D. Using dielectric grease in a connector

TASK E.5

Answer A is incorrect. Spread terminals are a common cause of voltage drop.

Answer B is incorrect. If the strands of a stranded wire get broken, then it reduces the size of the wire and causes a voltage drop.

Answer C is incorrect. Greenish corrosion known as *green death* is a common cause of voltage drop.

Answer D is correct. Dielectric grease is used in a connector to protect it from moisture, but will not cause a voltage drop.

TASK A.20

35. A vehicle is being diagnosed for a poor drivability complaint. The vehicle also fails to start at times; it just cranks with no-start. Technician A says the charging system could be undercharging at times. Technician B says an under- or overcharging system should generate a trouble code. Who is correct?

   A.   A only
   B.   B only
   C.   Both A and B
   D.   Neither A nor B

   Answer A is incorrect. Technician B is also correct.

   Answer B is incorrect. Technician A is also correct.

   Answer C is correct. Both Technicians are correct. An undercharged system would not keep the battery at a full state of charge, but should generate a diagnostic trouble code for over or undercharging.

   Answer D is incorrect. Both Technicians are correct.

TASK B.5

36. All of the following are true when testing a pickup coil EXCEPT:

   A.   Resistance value should fall within manufacturer's specifications.
   B.   Pickup coils should produce an AC voltage signal.
   C.   If adjustable, then the gap should be checked with a steel feeler gauge.
   D.   An erratic ohmmeter reading while wiggling the pickup coil wires indicates that the pickup coil has an intermittent open.

   Answer A is incorrect. The pickup coil resistance should fall with the manufacturer's specification, usually 500 to 1,500 ohms.

   Answer B is incorrect. Another method of testing the pickup coil is to measure the AC voltage signal.

   Answer C is correct. Some pickup coils are adjustable and should be adjusted using a brass feeler gauge. The magnet in the pickup coil will simulate a drag against the steel feeler gauge, giving a false reading.

   Answer D is incorrect. Any erratic reading when measuring the resistance of the pickup coil indicates an intermittent open in the pickup coil.

TASK C.3

37. Nylon fuel lines should be inspected for all of the following EXCEPT:

   A.   Discoloration
   B.   Loose fitting
   C.   Cracks
   D.   Kinks

   Answer A is correct. Nylon lines can become discolored over time. This discoloration does not affect the integrity of the line.

   Answer B is incorrect. A loose-fitting line will usually leak and should be repaired if found.

   Answer C is incorrect. Any noticeable crack in a nylon fuel line will eventually leak and should be replaced.

   Answer D is incorrect. Any kinks in nylon fuel lines will restrict fuel flow. Kinked lines should be replaced.

38. Technician A says that when checking a pulsed-air secondary air-injection system, exhaust pressure pulses felt at the fresh-air intake hose indicate a bad check or reed valve. Technician B says that when testing secondary air-injection systems, the gas analyzer can be used to confirm normal air-injection operation. Who is correct?

TASK D.3.3

A. A only

B. B only

C. Both A and B

D. Neither A nor B

Answer A is incorrect. Technician B is also correct.

Answer B is incorrect. Technician A is also correct.

Answer C is correct. Both Technicians are correct. The secondary air system is used to supply additional air to the catalytic converter. Some systems use an air pump, some use only check valves or reed valves, and some use both. Any exhaust felt at the check valve is an indication of a bad check valve. The gas analyzer can be used to verify secondary air-injection operation.

Answer D is incorrect. Both Technicians are correct.

39. Technician A says a digital voltmeter cannot be used to check an O_2 sensor. Technician B says a test light can be used to check an O_2 sensor. Who is correct?

TASK E.4

A. A only

B. B only

C. Both A and B

D. Neither A nor B

Answer A is incorrect. A digital voltmeter is used to monitor the voltage level on an oxygen sensor.

Answer B is incorrect. A test light cannot be used to test an oxygen sensor.

Answer C is incorrect. Neither Technician is correct.

Answer D is correct. Neither Technician is correct. When testing an oxygen sensor, a digital voltage meter or oscilloscope are the best choices. A scan tool can also be used.

40. All of the following are measured by a 5-gas analyzer EXCEPT:

TASK A.10

A. Oxides of nitrogen (NOx)

B. Smog

C. CO

D. HCs

Answer A is incorrect. NOx are measured with a 5-gas analyzer.

Answer B is correct. Smog is a photochemical mixture of gases that cannot be measured with a 5-gas analyzer.

Answer C is incorrect. Carbon monoxide can be measured with a 5-gas analyzer and is expressed in percentage.

Answer D is incorrect. Hydrocarbon is measured with a 5-gas analyzer and is expressed in parts per million.

TASK A.21

41. Technician A says the alternator connectors should always be inspected for corrosion and/or distortion from overheating when replacing an alternator. Technician B says some replacement alternators come with a new connector. Who is correct?

    A. A only
    B. B only
    C. Both A and B
    D. Neither A nor B

    Answer A is incorrect. Technician B is also correct.

    Answer B is incorrect. Technician A is also correct.

    Answer C is correct. Both Technicians are correct. Alternator connectors should always be inspected for corrosion and/or distortion from overheating. Some replacement alternators come with a new connector.

    Answer D is incorrect. Both Technicians are correct.

TASK B.5

42. Technician A says most 2-wire COP systems use individual ignition control modules for each cylinder. Technician B says a 4-wire COP coil is checked like any other ignition coil. Who is correct?

    A. A only
    B. B only
    C. Both A and B
    D. Neither A nor B

    Answer A is incorrect. Technician B is also correct.

    Answer B is incorrect. Technician A is also correct.

    Answer C is correct. Both Technicians are correct. There are basically two different types of COP systems. A 4-wire COP has the ignition module integrated with the coil. A 2-wire COP has a remote ignition control module. The coil on both types is tested like any other coil.

    Answer D is incorrect. Both Technicians are correct.

TASK C.1

43. An SFI vehicle stalls intermittently at idle and has negative long-term fuel trim correction values stored when checked with a scan tool. All of the following conditions could cause this EXCEPT:

    A. Leaking fuel injectors
    B. A leaking fuel pressure regulator diaphragm
    C. A vacuum leak
    D. A fuel pressure regulator stuck closed

    Answer A is incorrect. Leaking fuel injectors would cause a rich mixture and negative fuel trim.

    Answer B is incorrect. A leaking fuel pressure regulator diaphragm would cause a rich mixture and a negative fuel trim.

    Answer C is correct. A vacuum leak would cause a high idle and fuel trim to go positive not negative.

    Answer D is incorrect. A stuck closed fuel pressure regulator would cause a negative fuel trim.

44. Technician A says a stuck-open purge control valve could cause a hesitation on takeoff. Technician B says the purge control valve is a normally open solenoid. Who is correct?

    A. A only

    B. B only

    C. Both A and B

    D. Neither A nor B

    TASK D.4.1

    Answer A is correct. Only Technician A is correct. If the purge solenoid were to stick open, then the canister would be purging all the time causing the engine to run too rich. This could cause a hesitation.

    Answer B is incorrect. The purge control solenoid is a normally closed solenoid. The vent solenoid is the normally open solenoid.

    Answer C is incorrect. Only Technician A is correct.

    Answer D is incorrect. Technician A is correct.

45. An SFI engine has poor acceleration when the vehicle is suddenly accelerated to WOT. Idle and cruise performance is fine. Technician A says a faulty MAF sensor could cause this. Technician B says a weak fuel pump could cause this. Who is correct?

    A. A only

    B. B only

    C. Both A and B

    D. Neither A nor B

    TASK E.3

    Answer A is incorrect. A faulty MAF sensor would most likely cause problems under both cruise and WOT acceleration conditions.

    Answer B is correct. Only Technician B is correct. A weak fuel pump could cause poor acceleration when suddenly going to WOT.

    Answer C is incorrect. Only Technician B is correct.

    Answer D is incorrect. Technician B is correct.

46. All of the following are part of reprogramming the PCM EXCEPT:

    A. Connect a special type of battery charger to vehicle

    B. Connect scan tool to DLC (diagnostic link connector)

    C. Start the engine for 15 seconds

    D. Validate the vehicle VIN (vehicle identification number) number

    TASK E.7

    Answer A is incorrect. The battery should be fully charged. A battery charger can be used if it is the type that delivers a constant voltage with no fluctuations.

    Answer B is incorrect. The scan tool is connected to the DLC to transfer data.

    Answer C is correct. The reprogramming procedure is performed with the key on and the engine off.

    Answer D is incorrect. Validation of the VIN number is part of reprogramming the PCM.

TASK A.5

47. During a vacuum test the vacuum gauge shows a rapidly fluctuating motion from 15 to 21 in. Hg at idle. Technician A says this could be caused by a loose intake manifold. Technician B says this could be caused by a burned exhaust valve. Who is correct?

   A. A only
   B. B only
   C. Both A and B
   D. Neither A nor B

   Answer A is incorrect. A leaking intake manifold would cause a steady low vacuum reading.

   Answer B is correct. Only Technician B is correct. A burnt valve would cause the needle on the vacuum gauge to rapidly fluctuate between 15 and 21 in. Hg.

   Answer C is incorrect. Only Technician B is correct.

   Answer D is incorrect. Technician B is correct.

TASK E.4

48. Technician A says a faulty TPS can cause a dead spot in the throttle when accelerating. Technician B says faulty TPS will always set a DTC. Who is correct?

   A. A only
   B. B only
   C. Both A and B
   D. Neither A nor B

   Answer A is correct. Only Technician A is correct. As a TPS wears, it can develop an open spot in it were it rides the most. When driving a vehicle with a bad spot in the TPS, the vehicle will have a dead spot in the throttle when the bad portion of the TPS is reached.

   Answer B is incorrect. The TPS will not always set a DTC even if it is defective.

   Answer C is incorrect. Only Technician A is correct.

   Answer D is incorrect. Technician A is correct

TASK A.8

49. During a cylinder leak down test on a 6-cylinder engine, air is heard coming from the #2 spark plug hole as cylinder #3 is being checked. Technician A says that this could be caused by a blown head gasket. Technician B says this could be caused by a cracked engine block. Who is correct?

   A. A only
   B. B only
   C. Both A and B
   D. Neither A nor B

   Answer A is incorrect. Technician B is also correct.

   Answer B is incorrect. Technician A is also correct.

   Answer C is correct. Both Technicians are correct. During a leak down test, air is injected into the cylinder while the piston is at TDC of the compression stroke. If air is heard escaping from an adjacent cylinder, a blown head gasket or cracked head or block should be suspected.

   Answer D is incorrect. Both Technicians are correct.

50. Technician A says when high voltage drop is found in a circuit, check for burned wires, connector ring terminals, loose retaining nuts, or other wire and connector concerns. Technician B says that greenish white corrosion can happen at any point where the wire insulation has been pierced or opened in any way. Who is correct?

    A. A only
    B. B only
    C. Both A and B
    D. Neither A nor B

    Answer A is incorrect. Technician B is also correct.

    Answer B is incorrect. Technician A is also correct.

    Answer C is correct. Both Technicians are correct. Anything that increases the resistance in a circuit will increase the voltage drop in a circuit. When voltage drop is found, check for loose connections and loose splices. Also look for signs of greenish white corrosion.

    Answer D is incorrect. Both Technicians are correct.

PREPARATION EXAM 5—ANSWER KEY

1.	D	21.	A	41.	C
2.	A	22.	C	42.	A
3.	A	23.	C	43.	B
4.	D	24.	A	44.	C
5.	B	25.	B	45.	D
6.	B	26.	B	46.	B
7.	C	27.	A	47.	C
8.	D	28.	C	48.	B
9.	C	29.	C	49.	A
10.	B	30.	D	50.	A
11.	C	31.	A		
12.	A	32.	A		
13.	D	33.	A		
14.	D	34.	B		
15.	C	35.	D		
16.	D	36.	A		
17.	C	37.	D		
18.	C	38.	B		
19.	C	39.	B		
20.	C	40.	A		

PREPARATION EXAM 5—EXPLANATIONS

TASK A.14

1. With a thermometer taped to the upper radiator hose and the vehicle fully warmed up, the temperature indication should be which of the following?

    A. Lower than the temperature measured at the lower radiator hose

    B. One-third of the engine thermostat temperature rating

    C. Ambient temperature

    D. Within a few degrees of the engine thermostat temperature rating

    Answer A is incorrect. The upper radiator hose will measure a higher temperature than the lower because the coolant has not traveled through the radiator releasing the heat it has received.

    Answer B is incorrect. The temperature would be much higher than one-third of the engine thermostat temperature rating.

    Answer C is incorrect. Ambient temperature is the outside temperature; the upper radiator hose would be much hotter than that.

    Answer D is correct. On a fully warmed up vehicle, the upper radiator hose should measure within a few degrees of the engine thermostat temperature rating.

TASK B.1

2. Technician A says timing specifications can be found on the under hood vehicle emissions control information (VECI) sticker. Technician B says the late ignition timing cause the engine to crank slow on a warm engine. Who is correct?

    A. A only

    B. B only

    C. Both A and B

    D. Neither A nor B

    Answer A is correct. Only Technician A is correct. The ignition timing specification can be found on the VECI.

    Answer B is incorrect. Late (retarded) ignition timing will cause low power complaints, early (advanced) ignition timing will cause slow cranking or engine baulking when cranking.

    Answer C is incorrect. Only Technician A is correct.

    Answer D is incorrect. Technician A is correct.

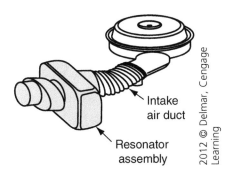

Intake air duct

Resonator assembly

2012 © Delmar, Cengage Learning

3. Refer to the above illustration. Technician A says the resonator assembly is used to reduce noise. Technician B says a leak in the intake air duct would allow unfiltered air to enter the engine. Who is correct?

A. A only
B. B only
C. Both A and B
D. Neither A nor B

TASK C.9

Answer A is correct. Only Technician A is correct. Resonators are sometimes used on the intake air system to lower the noise produced by the incoming air.

Answer B is incorrect. A leak in the intake air duct would produce noise, but the air entering through the leak would still pass through the air filter before entering the engine.

Answer C is incorrect. Only Technician A is correct.

Answer D is incorrect. Technician A is correct.

4. All of the following are components of a PCV system EXCEPT:

A. PCV heater
B. Valve grommet
C. Inlet filter
D. Purge control solenoid

TASK D.1.2

Answer A is incorrect. Many late model vehicles now use a PCV heater to prevent restriction of the valve in cold weather.

Answer B is incorrect. The PCV valve on many vehicles goes in a grommet located in the valve cover.

Answer C is incorrect. Some PCV systems use a separate inlet filter; others use the main air filter to filter the PCV inlet.

Answer D is correct. The purge control solenoid is part of the EVAP system, not the PCV system.

5. Technician A says continuous monitors consist of catalyst efficiency, misfire, and component monitors. Technician B says the continuous monitors consist of three separate monitors. Who is correct?

A. A only
B. B only
C. Both A and B
D. Neither A nor B

TASK E.1

Answer A is incorrect. The three continuous monitors are misfire, fuel system, and comprehensive component monitors.

Answer B is correct. Only Technician B is correct. There are three separate continuous monitors.

Answer C is incorrect. Only Technician B is correct.

Answer D is incorrect. Technician B is correct.

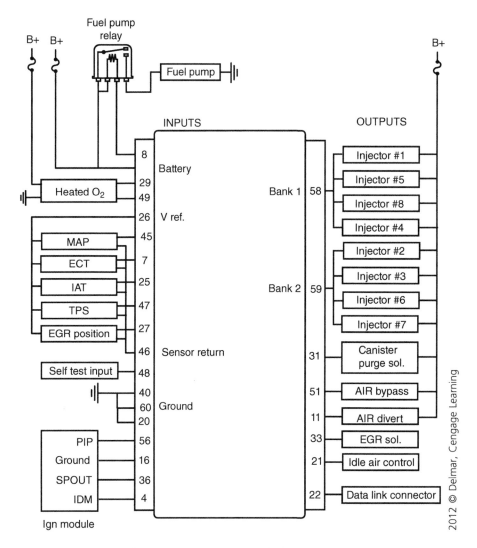

TASK C.1

6. Refer to the above illustration. A multi-port, fuel-injected distributor-type ignition V8 engine is running rough and has a lean fuel mixture. The fuel injectors are pair-fired in two groups consisting of cylinders #2, #3, #6 and #7 and cylinders #1, #4, #5 and #8. Injecting propane will smooth out engine idle quality. There are no vacuum leaks. After performing a cylinder balance test, cylinders #2, #3 and #6 are found to be weak. Technician A says a bad ignition module could cause this. Technician B says the low resistance on cylinder #2 fuel injector could cause this. Who is correct?

A. A only

B. B only

C. Both A and B

D. Neither A nor B

Answer A is incorrect. An ignition module on a distributor-type ignition system would affect all cylinders, not just three.

Answer B is correct. Only Technician B is correct. If fuel injector #2 were to have low resistance, then the injector would use more current and would not leave enough current for the other injectors in the group.

Answer C is incorrect. Only Technician B is correct.

Answer D is incorrect. Technician B is correct.

7. Technical service bulletins should be used for all of the following EXCEPT:

   A. Updated service manual procedures

   B. Updated repair procedures for certain complaints

   C. Updated vehicle recalls

   D. Updated flash programming

TASK A.2

   Answer A is incorrect. When a service manual update is issued, a TSB (technical service bulletins) is issued addressing the change.

   Answer B is incorrect. As changes are made to certain repair procedures, a TSB will be issued.

   Answer C is correct. A TSB is not issued for a vehicle recall, a vehicle recall is issued.

   Answer D is incorrect. If a new updated flash program is issued to fix a certain complaint, then a TSB will be issued.

8. All of the following are inputs for electronic timing control EXCEPT:

   A. Knock sensor

   B. Throttle position sensor (TPS)

   C. Engine coolant temperature

   D. Oxygen sensor

TASK B.5

   Answer A is incorrect. The knock sensor detect spark knock and sends this input to the PCM (powertrain control module) for timing retarding.

   Answer B is incorrect. The throttle position is used to determine spark advance.

   Answer C is incorrect. The engine coolant temperature is used to determine the timing control.

   Answer D is correct. The oxygen sensor is not part of the timing control inputs.

9. Technician A says that if some exhaust gas recirculation (EGR) passages are plugged in the intake manifold, then the engine may have detonation when accelerating. Technician B says that plugged EGR passages may cause a failed IM240 emissions test for high oxides of nitrogen (NOx). Who is correct?

TASK D.2.1

   A. A only

   B. B only

   C. Both A and B

   D. Neither A nor B

   Answer A is incorrect. Technician B is also correct.

   Answer B is incorrect. Technician A is also correct.

   Answer C is correct. Both Technicians are correct. The purpose of the EGR system is to lower the combustion chamber temperature to lower NOx formation in the engine. Any restriction in the EGR system can allow the combustion chamber to overheat causing detonation and the increased formation of NOx.

   Answer D is incorrect. Both Technicians are correct.

TASK E.4

10. All of the following types of test equipment are used to check the O_2 sensor EXCEPT:

    A. Lab scope
    B. Low-impedance volt meter
    C. Graphing meter
    D. High-impedance volt meter

    Answer A is incorrect. The lab scope is a vital tool for testing the O_2 sensor operation.

    Answer B is correct. A low-impedance volt meter draws an excessive amount of amperage and should not be used to test electronic engine control sensors.

    Answer C is incorrect. A graphing meter is a very useful tool when testing O_2 sensors.

    Answer D is incorrect. A high-impedance voltmeter can be used to test a O_2 sensor.

TASK A.6

11. During a cylinder power balance test, there is no RPM drop on cylinder #4. Technician A says that the cylinder is not contributing to the power flow of the engine. Technician B says that the cylinder may have an inoperative spark plug. Who is correct?

    A. A only
    B. B only
    C. Both A and B
    D. Neither A nor B

    Answer A is incorrect. Technician B is also correct.

    Answer B is incorrect. Technician A is also correct.

    Answer C is correct. Both Technicians are correct. During a power balance test, each cylinder's spark is shorted to ground one at a time, and the RPM drop is noted. Any cylinder that does not drop in RPM when shorted is not contributing to the power flow of the engine. Injectors, spark plugs and wires, and mechanical components can fail and cause a cylinder not to contribute.

    Answer D is incorrect. Both Technicians are correct.

TASK B.1

12. A vehicle with electronic spark timing control has low power and needs its ignition timing adjusted. Technician A says ignition timing on a distributor-type ignition is usually set with the engine running and the computer control circuit disconnected or disabled. Technician B says ignition timing is usually set with the engine running at 2,500 rpm. Who is correct?

    A. A only
    B. B only
    C. Both A and B
    D. Neither A nor B

    Answer A is correct. Only Technician A is correct. The computer is in control of the ignition timing and this control must be disabled before the ignition timing can be adjusted.

    Answer B is incorrect. The ignition timing is adjusted at idle speed; the timing advance can be checked at 2,500 rpm.

    Answer C is incorrect. Only Technician A is correct.

    Answer D is incorrect. Technician A is correct.

13. A vehicle is being diagnosed for a PO134 trouble code. (Oxygen Sensor Circuit, No Activity Detected) (bank #1, sensor #1) with a bank #1 fuel trim of +20. Technician A says the problem could be high fuel system pressure. Technician B says the problem is more likely confined to the downstream oxygen sensor on the bank were cylinder #1 is on. Who is correct?

TASK C.2

   A. A only
   B. B only
   C. Both A and B
   D. Neither A nor B

   Answer A is incorrect. A high fuel system pressure would affect all O_2 sensors and would cause a negative fuel trim, not a positive fuel trim.

   Answer B is incorrect. The downstream O_2 sensor on the bank with cylinder #1 would be bank #1, sensor #2.

   Answer C is incorrect. Neither Technician is correct.

   Answer D is correct. Neither Technician is correct. When diagnosing drivability complaints with O_2 sensor readings and fuel trim. Bank #1 is the bank with cylinder #1 and sensor #1 is the upstream sensor. Fuel trim is the adding and subtracting of fuel for mixture correction a+ means fuel is being added, and a– means fuel is being subtracted.

14. The pulse air-injection system is driven by which of the following?

TASK D.3.1

   A. A drive belt
   B. A helical gear
   C. An electric motor
   D. Negative pressure pulses in the exhaust.

   Answer A is incorrect. Air pumps can be belt driven, but not pulse air-injection systems.

   Answer B is incorrect. Pulse air-injection systems are not gear driven.

   Answer C is incorrect. Air pumps can be electrically driven, but not pulse air-injection systems.

   Answer D is correct. The pulse air-injection works from the negative pulses from the exhaust.

15. Technician A says a voltage drop test checks the amount of resistance between two test points in a circuit. Technician B says more than a 0.5 volts drop indicates excessive resistance across the battery positive cable. Who is correct?

TASK E.5

   A. A only
   B. B only
   C. Both A and B
   D. Neither A nor B

   Answer A is incorrect. Technician B is also correct.

   Answer B is incorrect. Technician A is also correct.

   Answer C is correct. Both Technicians are correct. A voltage drop test is one of the best ways to determine if a circuit has high resistance. A reading of 0.5 volts when doing a voltage drop test is an indication of high resistance in the circuit.

   Answer D is incorrect. Both Technicians are correct.

TASK A.14

16. A proper cooling system inspection involves all of the following EXCEPT:

    A.   A thermostat operation inspection

    B.   A coolant condition inspection

    C.   A cooling fan operation inspection

    D.   A condenser inspection

Answer A is incorrect. The thermostat should be tested for proper operation during a cooling system inspection.

Answer B is incorrect. The coolant should be inspected for condition and color.

Answer C is incorrect. The fan should be tested for proper operation during a cooling system inspection.

Answer D is correct. The condenser is part of the air conditioning system, not the cooling system.

TASK A.10

17. High carbon monoxide (CO) emissions may be caused by all the following EXCEPT:

    A.   Rich air/fuel mixture

    B.   Exhaust manifold leak

    C.   Fouled spark plug

    D.   Malfunctioning secondary air switching valve

Answer A is incorrect. A rich air/fuel mixture will cause high CO readings.

Answer B is incorrect. An exhaust leak will allow outside oxygen to be drawn in where the oxygen sensor will read the extra oxygen and send that information to the PCM. The PCM will command a richer mixture, causing an increase in CO.

Answer C is correct. CO is the result from incomplete combustion; a fouled spark plug would produce no combustion, so the formation of CO would not occur.

Answer D is incorrect. If the air-switching valve fails to switch the secondary air downstream, then the oxygen sensor will read the extra oxygen and send that information to the PCM. The PCM will command a richer mixture, causing an increase in CO.

TASK B.1

18. A vehicle with an electronic ignition fails to start. Technician A says this could be caused by a defective cam shaft sensor connection. Technician B says this could be caused by an ignition coil. Who is correct?

    A.   A only

    B.   B only

    C.   Both A and B

    D.   Neither A nor B

Answer A is incorrect. Technician B is also correct.

Answer B is incorrect. Technician A is also correct.

Answer C is correct. Both Technicians are correct. While some vehicles will still start without the cam sensor signal many will not, so the cam sensor connection could cause a no-start on some vehicles. A failed ignition coil will cause a no-start on vehicles using only one ignition coil. On vehicles using multiple coils a failed coil will cause a miss.

Answer D is incorrect. Both Technicians are correct.

19. Technician A says an enhanced EVAP system must be able to detect a leak as small as 0.020 inches. Technician B says the enhanced system must be able to detect a loose gas cap. Who is correct?

TASK D.4.1

   A. A only

   B. B only

   C. Both A and B

   D. Neither A nor B

Answer A is incorrect. Technician B is also correct.

Answer B is incorrect. Technician A is also correct.

Answer C is correct. Both Technicians are correct. An enhanced EVAP system can test the integrity of the EVAP system. A leak as small as 0.020" can be detected on a vehicle with enhanced EVAP. A loose gas cap is the cause of many malfunctioning indicator lamps being illuminated.

Answer D is incorrect. Both Technicians are correct.

20. A PCM is being replaced. Technician A says the new PCM should be ordered using the original PCM's part number. Technician B says installing a used PCM can cause the theft deterrent system to activate on some vehicles. Who is correct?

TASK E.7

   A. A only

   B. B only

   C. Both A and B

   D. Neither A nor B

Answer A is incorrect. Technician B is also correct.

Answer B is incorrect. Technician A is also correct.

Answer C is correct. Both Technicians are correct. With all the different configurations of engine operating system care must be taken when replacing the PCM. The best place to start is with the original PCM part number. If a salvage yard PCM is being installed, then the part numbers must match; even if they do, sometimes the theft deterrent system will activate on some vehicles.

Answer D is incorrect. Both Technicians are correct.

21. While testing a turbocharger, the maximum boost pressure observed is 4 psi (27.6 kPa), while the specified pressure is 9 psi (62 kPa). Technician A says the engine compression may be low. Technician B says the wastegate may be sticking closed. Who is correct?

TASK C.14

   A. A only

   B. B only

   C. Both A and B

   D. Neither A nor B

Answer A is correct. Only Technician A is correct. Low engine compression will affect the amount of boost produced by the turbocharger.

Answer B is incorrect. If the wastegate were to stick closed, then it would cause an over-boost problem, not an under-boost problem.

Answer C is incorrect. Only Technician A is correct.

Answer D is incorrect. Technician A is correct.

TASK E.7

22. A vehicle has had its PCM replaced for a charging system problem; now the vehicle will not start. Technician A says the replacement PCM on some vehicles require being flash-programmed in order to operate. Technician B says the flash program for some scan tools can be downloaded and installed after the PCM is in the vehicle. Who is correct?

   A. A only
   B. B only
   C. Both A and B
   D. Neither A nor B

   Answer A is incorrect. Technician B is also correct.

   Answer B is incorrect. Technician A is also correct.

   Answer C is correct. Both Technicians are correct. Most late model vehicles require the PCM be flash programmed when a new PCM is installed. Some scan tools allow the flash to be downloaded in the scan tool and installed after the PCM is in the vehicle.

   Answer D is incorrect. Both Technicians are correct.

TASK A.12

23. A timing belt that has jumped one tooth can cause all EXCEPT:

   A. Low power
   B. Poor fuel economy
   C. High engine idle
   D. Poor vehicle stopping

   Answer A is incorrect. A timing belt that is off one tooth will cause a lack of engine power.

   Answer B is incorrect. Poor fuel economy is one of the complaints from the owner of a vehicle with a jumped timing belt.

   Answer C is correct. A jumped timing belt will cause a low idle, not a high idle.

   Answer D is incorrect. A jumped timing belt will cause low engine vacuum, which prevents the brake booster from doing its job and causes poor vehicle braking.

TASK B.1

24. Which of the following conditions would most likely cause weak spark at all the spark plug wires?

   A. High primary circuit resistance
   B. Low secondary resistance
   C. Secondary spark plug wire insulation breakdown
   D. Ignition timing out of adjustment

   Answer A is correct. High primary resistance will cause the secondary voltage that is produced to be low or weak.

   Answer B is incorrect. Low secondary resistance is a good thing and will not cause a weak spark.

   Answer C is incorrect. While secondary insulation will affect one or two cylinders, it will not affect all of them.

   Answer D is incorrect. Ignition timing out of adjustment will cause drivability problems, it will not cause weak spark.

25. Technician A says a leaking injector could cause a high CO_2 (carbon dioxide) reading at idle. Technician B says that no vacuum to the pressure regulator could cause a high CO reading at idle. Who is correct?

TASK C.1

   A. A only

   B. B only

   C. Both A and B

   D. Neither A nor B

   Answer A is incorrect. A leaking injector would cause a high CO reading because of the rich air/fuel mixture, but would not cause a high CO_2 reading. The CO_2 would be lowered if the injector was leaking.

   Answer B is correct. Only Technician B is correct. With the vacuum signal absent from the fuel pressure regulator the fuel pressure would be about 10 psi higher than normal, causing a rich air/fuel mixture and a high CO reading.

   Answer C is incorrect. Only Technician B is correct.

   Answer D is incorrect. Technician B is correct.

26. Technician A says if the EGR valve remains open at idle and low speed, then the idle will be high. Technician B says if the EGR valve does not open at cruising speeds, then detonation can occur. Who is correct?

TASK D.2.1

   A. A only

   B. B only

   C. Both A and B

   D. Neither A nor B

   Answer A is incorrect. If the EGR valve opens at idle or low speeds, then the engine will run rough and possibly stall.

   Answer B is correct. Only Technician B is correct. The EGR system is used to lower combustion chamber temperature. If the EGR valve does not open, then the high-combustion chamber temperature can cause detonation to occur.

   Answer C is incorrect. Only Technician B is correct.

   Answer D is incorrect. Technician B is correct.

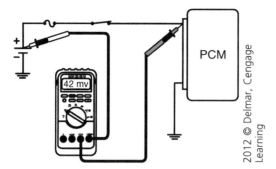

TASK E.5

27. Technician A says that when measuring voltage drop as shown in the illustration, it must be done with a digital multimeter (DMM) and the circuit that is being tested must be operating. Technician B says a high voltage drop reading indicates low resistance in the circuit. Who is correct?

   A.   A only
   B.   B only
   C.   Both A and B
   D.   Neither A nor B

   Answer A is correct. Only Technician A is correct. A voltage drop test is performed to determine if resistance is in the circuit. A DMM is connected to the two points between which the voltage drop is to be measured. The circuit must be turned on.

   Answer B is incorrect. A high voltage drop reading indicates high resistance in the circuit.

   Answer C is incorrect. Only Technician A is correct.

   Answer D is incorrect. Technician A is correct.

TASK E.9

28. Technician A says an OBD II (on-board diagnostics second generation) warm-up cycle is defined as a trip in which the engine temperature increases by at least 40°F (22°C) and reaches 160°F (70°C) during one key on and the engine running (KOER) cycle. Technician B says a drive cycle refers to a trip in which the operating parameters required for the PCM to run OBD II emissions-related monitors has been met. Who is correct?

   A.   A only
   B.   B only
   C.   Both A and B
   D.   Neither A nor B

   Answer A is incorrect. Technician B is also correct.

   Answer B is incorrect. Technician A is also correct.

   Answer C is correct. Both Technicians are correct. For a warm up cycle to occur, the temperature must increase by at least 40°F (22°C) and reaches 160°F (70°C) during KOER cycle. Monitors will not be run until certain conditions are met during a drive cycle.

   Answer D is incorrect. Both Technicians are correct.

29. A restricted exhaust will cause vacuum readings to do what?

    A. Be 2 to 3 inches higher than normal
    B. Be 2 to 3 inches lower than normal
    C. Gradually drop as engine speed is increased
    D. Fluctuate between 17 and 20 inches

TASK A.5

    Answer A is incorrect. A high vacuum reading is good.

    Answer B is incorrect. If vacuum is a little low, then suspect a vacuum leak or late timing.

    Answer C is correct. A restricted exhaust will show a gradual decrease in vacuum as the engine speed is increased.

    Answer D is incorrect. A fluctuating vacuum reading is an indication of a burnt valve or broken valve spring.

30. A charging system that is overcharging can be caused by all of the following EXCEPT:

    A. An internally shorted battery
    B. Shorted field windings
    C. A defective voltage regulator
    D. High generator field winding circuit resistance

TASK A.20

    Answer A is incorrect. An internally shorted battery will cause an overcharging condition.

    Answer B is incorrect. Shorted field winding will cause an overcharging condition.

    Answer C is incorrect. A defective voltage regulator will cause an overcharging condition.

    Answer D is correct. High generator field winding resistance would cause an undercharging condition, not an overcharge.

31. An engine equipped with a distributorless ignition system (DIS) will not start. Technician A says a defective crankshaft position sensor could cause this. Technician B says a shorted ground wire to the DIS assembly could be the cause. Who is correct?

    A. A only
    B. B only
    C. Both A and B
    D. Neither A nor B

TASK B.1

    Answer A is correct. Only Technician A is correct. A defective crankshaft position sensor will cause a no-start condition.

    Answer B is incorrect. A shorted ground wire will not cause a no-start because the ground wire is already grounded.

    Answer C is incorrect. Only Technician A is correct.

    Answer D is incorrect. Technician A is correct.

TASK C.1

32. A vehicle with a lean exhaust code is being diagnosed. Technician A says the fuel pressure should be checked. Technician B says remove the vacuum hose from the fuel pressure regulator and check for fuel at the regulator vacuum nipple. Who is correct?

   A. A only

   B. B only

   C. Both A and B

   D. Neither A nor B

   Answer A is correct. Only Technician A is correct. Any time a lean or rich exhaust code is being diagnosed the fuel pressure should be tested. Low fuel pressure can cause a lean exhaust code.

   Answer B is incorrect. A leaking fuel pressure regulator diaphragm would cause fuel to be drawn into the intake manifold, richening the fuel mixture, not leaning it.

   Answer C is incorrect. Only Technician A is correct.

   Answer D is incorrect. Technician A is correct.

TASK D.4.1

33. Which valves are normally closed on an evaporative emissions control system?

   A. Canister purge valve

   B. Both canister purge and canister vent valve

   C. Canister vent valve

   D. Neither canister purge nor canister vent valve

   Answer A is correct. The canister purge valve is normally closed and opened by the PCM (powertrain control module) when canister purge is needed.

   Answer B is incorrect. Only the canister purge valve is normally closed.

   Answer C is incorrect. The canister vent valve is normally open and closed by the PCM when an integrity check is being made on the EVAP system.

   Answer D is incorrect. Only the canister purge valve is normally closed.

34. The PCM will automatically clear a Type B DTC (diagnostic trouble code) if there are no additional faults detected after which of the following?

TASK E.1

   A. Eighty warm-up cycles

   B. Forty consecutive warm-up cycles

   C. Two consecutive trips

   D. Four key-on/key-off cycles

   Answer A is incorrect. A Type B code is cleared after 40 warm-up cycles. A Type A code takes 80 warm-up cycles.

   Answer B is correct After 40 consecutive warm-up cycles with no further faults detected, the PCM will automatically clear the DTC on a Type B code.

   Answer C is incorrect. Three consecutive trips without a fault will disable the MIL.

   Answer D is incorrect. Four key on/key off cycles will do nothing.

35. A vehicle with a SFI (sequential fuel injection) V6 engine and OBD II emissions controls has set a DTC PO172 (System Too Rich, Bank 1). No other drivability concerns are present. The freeze frame data shows the code was set under warm idle conditions. Technician A says the problem could be an intake manifold vacuum leak. Technician B says the problem could be a defective fuel pressure regulator. Who is correct?

TASK E.2

    A. A only
    B. B only
    C. Both A and B
    D. Neither A nor B

    Answer A is incorrect. A leaking intake manifold would cause a lean code, not a rich code.

    Answer B is incorrect. A defective fuel pressure regulator would affect both engine banks, not just one.

    Answer C is incorrect. Neither Technician is correct.

    Answer D is correct. Neither Technician is correct. When diagnosing a bank-related trouble code, a knowledge of the operating system is crucial for proper diagnosis.

36. A vacuum leak is the suspected cause for a rough idle concern. Using a 5-gas analyzer, Technician A says O_2 will be higher than normal if a vacuum leak is present. Technician B says CO will be higher than normal if a vacuum leak is present. Who is correct?

TASK A.10

    A. A only
    B. B only
    C. Both A and B
    D. Neither A nor B

    Answer A is correct. Only Technician A is correct. Any additional oxygen entering the engine will cause the O_2 readings on the 5-gas analyzer to be higher than normal.

    Answer B is incorrect. A vacuum leak would cause the CO to be lower, not higher.

    Answer C is incorrect. Only Technician A is correct.

    Answer D is incorrect. Technician A is correct.

37. To check ignition coil available voltage output, the technician should do which of the following?

TASK B.1

    A. Disconnect the plug wire at the plug and ground it
    B. Disconnect the fuel pump power lead
    C. Disconnect the coil wire and ground it
    D. Conduct the test using a suitable spark tester that requires 25 kV

    Answer A is incorrect. Grounding a plug wire will not show available voltage.

    Answer B is incorrect. Disconnecting the fuel pump is not necessary when checking available voltage.

    Answer C is incorrect. Grounding the coil wire would not show available voltage.

    Answer D is correct. A spark tester should be used to prevent voltage spikes. A 25 kV spark tester is sufficient for most electronic ignition systems.

TASK C.14

38. A turbocharged engine is experiencing excessive oil consumption and blue smoke from the tailpipe at idle and cruising speeds. This could be caused by which of the following?

   A. Dirty air filter
   B. A plugged oil return passage
   C. Restricted exhaust
   D. Turbo spinning too fast

   Answer A is incorrect. A dirty air filter will not cause the vehicle to have blue smoke from the exhaust.

   Answer B is correct. A plugged oil return passage can cause the oil to be forced past the turbocharger bearings.

   Answer C is incorrect. A restricted exhaust will not cause blue smoke and oil consumption.

   Answer D is incorrect. The turbo spins fast all the time. This will not cause turbo failure.

TASK D.3.1

39. Technician A says if the catalytic converter is restricted, then the engine will produce higher than normal vacuum at 2,000 rpm. Technician B says if you tap on a monolithic (honeycomb) converter with a rubber hammer and the converter rattles, then it should be replaced. Who is correct?

   A. A only
   B. B only
   C. Both A and B
   D. Neither A nor B

   Answer A is incorrect. A restricted catalytic converter will cause the vacuum to gradually decrease as the engine speed is increased.

   Answer B is correct. Only Technician B is correct. One test that can be done on a catalytic converter is a tap test. Using a rubber mallet, tap the converter and listen for any rattling. If rattling is heard, then the monolithic catalyst material is broken and the converter should be replaced.

   Answer C is incorrect. Only Technician B is correct.

   Answer D is incorrect. Technician B is correct.

TASK E.4

40. Technician A says the engine coolant temperature sensor voltage drop decreases as the coolant temperature increases. Technician A says the engine coolant temperature sensor resistance increases as the coolant temperature increases. Who is correct?

   A. A only
   B. B only
   C. Both A and B
   D. Neither A nor B

   Answer A is correct. Only Technician A is correct. As the temperature goes up, the resistance goes down on a coolant temperature sensor. If the resistance goes down as the temperature increases, then the voltage drop would decrease.

   Answer B is incorrect. Coolant temperature sensors used on automobiles are negative temperature coefficient sensors; meaning as temperature goes up resistance goes down.

   Answer C is incorrect. Only Technician A is correct.

   Answer D is incorrect. Technician A is correct.

41. Technician A says incorrect cam shaft timing can cause an engine not to start. Technician B says incorrect cam shaft timing may cause a power loss. Who is correct?

    A. A only
    B. B only
    C. Both A and B
    D. Neither A nor B

    TASK A.12

    Answer A is incorrect. Technician B is also correct.

    Answer B is incorrect. Technician A is also correct.

    Answer C is correct. Both Technicians are correct. If the cam shaft timing is incorrect, then the valves will open too early or too late. If the timing is off enough, then it can cause a no-start condition. If the cam shaft timing is out only a tooth or two, then the engine might start but would have low power.

    Answer D is incorrect. Both Technicians are correct.

42. Technician A says a scan tool can be used in conjunction with a thermometer to check thermostat operation. Technician B says thermostat operation can also be checked visually by running the engine from a cold start until it is hot. Who is correct?

    A. A only
    B. B only
    C. Both A and B
    D. Neither A nor B

    TASK A.14

    Answer A is correct. Only Technician A is correct. Using a scan tool along with a thermometer is a very accurate way to diagnosis thermostat operation.

    Answer B is incorrect. A thermostat cannot be accurately diagnosed by just running the vehicle cold to hot.

    Answer C is incorrect. Only Technician A is correct.

    Answer D is incorrect. Technician A is correct.

43. Spark plug wires are being measured for resistance. Technician A says if the ohmmeter reads OL, then the circuit has little or no resistance. Technician B says plug wires should measure about 5,000 to 10,000 ohms per foot. Who is correct?

    A. A only
    B. B only
    C. Both A and B
    D. Neither A nor B

    TASK B.4

    Answer A is incorrect. If the ohmmeter read OL, then there is too much resistance to be measured or out of limits.

    Answer B is correct. Only Technician B is correct. When testing spark plug wire resistance allow about 5,000 to 10,000 ohms per foot up to 30,000 ohms.

    Answer C is incorrect. Only Technician B is correct.

    Answer D is incorrect. Technician B is correct.

TASK C.1

44. Which of the following is the least likely cause of poor fuel mileage on a vehicle with SFI?

   A.   Plugged vacuum hose to the fuel pressure regulator
   B.   Plugged return fuel line
   C.   Plugged fuel filter
   D.   Defective thermostat

   Answer A is incorrect. A plugged vacuum hose to the regulator would cause the fuel pressure to be too high causing poor fuel mileage.

   Answer B is incorrect. If the return fuel line was plugged, then the vehicle would have excessive fuel pressure, causing poor fuel mileage.

   Answer C is correct. A plugged fuel filter would cause a low-power complaint, not poor fuel mileage.

   Answer D is incorrect. If the thermostat is defective and not allowing full operating temperature to be reached, then poor fuel mileage can be the result.

TASK D.4.3

45. All of the following are part of the evaporative emissions system EXCEPT:

   A.   Roll-over valve
   B.   Vapor canister
   C.   Gas cap
   D.   Pulse air feeder

   Answer A is incorrect. The roll-over valve is what the main vent hose connects to at the fuel tank. The roll-over valve closes in the event of a roll over.

   Answer B is incorrect. The vapor canister is filled with charcoal that absorbs the hydrocarbons until purged to the engine's intake system.

   Answer C is incorrect. The gas cap has a built-in vent valve and pressure-release valve.

   Answer D is correct. The pulse air feeder is part of the secondary air system.

TASK E.1

46. Technician A says many different adapters are needed if a scan tool is going to be used to retrieve DTCs from different manufactured ODB II (on-board diagnostics second generation) vehicles. Technician B says many different adapters are needed if a scan tool is going to be used to retrieve diagnostic trouble codes from different manufactured ODB I vehicles. Who is correct?

   A.   A only
   B.   B only
   C.   Both A and B
   D.   Neither A nor B

   Answer A is incorrect. OBD II vehicles all use the same 16-pin data link connector (DLC).

   Answer B is correct. Only Technician B is correct. OBD I vehicles all used a different DLC from each manufacture. This required many different adapters.

   Answer C is incorrect. Only Technician B is correct.

   Answer D is incorrect. Technician B is correct.

47. A vehicle has excessively high hydrocarbon (HC) emissions with no DTCs. Technician A says weak cylinder compression could be the cause. Technician B says a vacuum leak could be the cause. Who is correct?

TASK E.3

   A. A only
   B. B only
   C. Both A and B
   D. Neither A nor B

Answer A is incorrect. Technician B is also correct.

Answer B is incorrect. Technician A is also correct.

Answer C is correct. Both Technicians are correct. HC emissions are unburned fuel in the exhaust; weak compression, a burnt valve, and vacuum leaks are all causes of high hydrocarbon emissions.

Answer D is incorrect. Both Technicians are correct.

48. Air is escaping from the PCV valve opening in the valve cover during a cylinder leakage test. Technician A says a blown head gasket is a possible cause. Technician B says air escaping from the PCV valve opening in the valve cover is leaking past the rings. Who is correct?

TASK A.8

   A. A only
   B. B only
   C. Both A and B
   D. Neither A nor B

Answer A is incorrect. A blown head gasket would cause air to leak into the cooling system or adjacent cylinder.

Answer B is correct. Only Technician B is correct. Any air that leaks past the rings will enter the crankcase and exit out the PCV valve hole in the valve cover.

Answer C is incorrect. Only Technician B is correct.

Answer D is incorrect. Technician B is correct.

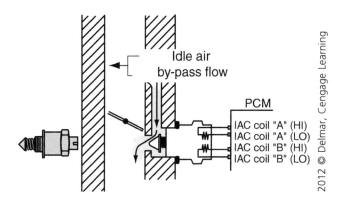

TASK C.1

49. A vehicle with the illustrated idle control system has a high idle. Technician A says a break in any of the wires can cause a high idle depending on the position of the plunger at the time of the break. Technician B says coil A high and coil B high are for high idle, and coil A low and coil B low are for low idle. Who is correct?

A. A only

B. B only

C. both A and B

D. Neither A nor B

Answer A is correct. Only Technician A is correct. The illustration shows a stepper-motor-type idle control system. A break in any of the four wires will prevent the idle air control motor from operating past the coil with the break in it. Idle speed will be whatever it was at the time of the break depending on plunger position.

Answer B is incorrect. Coil A high, Coil B high, Coil A low and Coil B low all refer to driver identification in the PCM.

Answer C is incorrect. Only Technician A is correct.

Answer D is incorrect. Technician A is correct.

TASK E.4

50. Technician A says a bi-directional scan tool is needed to actuate outputs such as the injectors for diagnostic purposes. Technician B says a bi-directional scan tool is needed to retrieve diagnostic trouble codes on an OBD II vehicle. Who is correct?

A. A only

B. B only

C. Both A and B

D. Neither A nor B

Answer A is correct. Only Technician A is correct. A bi-directional scan tool allows the technician to control actuators when diagnosing a vehicle.

Answer B is incorrect. A bi-directional scan tool is not needed to retrieve DTCs or scan tool data. Some tools are available that just read the data, but are unable to issue commands to the PCM.

Answer C is incorrect. Only Technician A is correct.

Answer D is incorrect. Technician A is correct.

PREPARATION EXAM 6—ANSWER KEY

1.	C	21.	B	41.	C
2.	A	22.	C	42.	A
3.	C	23.	B	43.	C
4.	A	24.	C	44.	D
5.	C	25.	D	45.	D
6.	C	26.	B	46.	C
7.	C	27.	B	47.	D
8.	A	28.	C	48.	B
9.	C	29.	C	49.	D
10.	D	30.	B	50.	B
11.	C	31.	B		
12.	B	32.	A		
13.	B	33.	D		
14.	A	34.	C		
15.	C	35.	C		
16.	D	36.	C		
17.	D	37.	D		
18.	B	38.	B		
19.	C	39.	A		
20.	D	40.	C		

PREPARATION EXAM 6—EXPLANATIONS

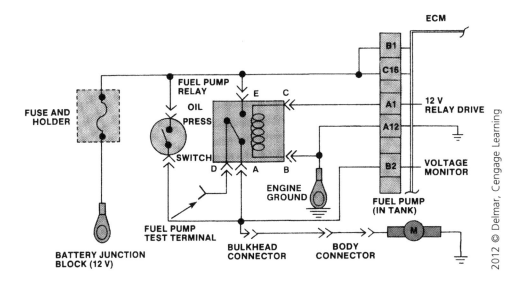

TASK A.16

1. Refer to the above illustration. Technician A says the double arrow (>>) indicates a male and female connector. Technician B says if there is only one arrow (>), then the connector does not connect with another connector. Who is correct?

    A. A only

    B. B only

    C. Both A and B

    D. Neither A nor B

    Answer A is incorrect. Technician B is also correct.

    Answer B is incorrect. Technician A is also correct.

    Answer C is correct. Both Technicians are correct. Connectors are identified by a double or single arrow. If it is a double arrow, then one connector connects with another connector or component. If the symbol is a single connector, then the connector does not connect with another connector or component.

    Answer D is incorrect. Both Technicians are correct.

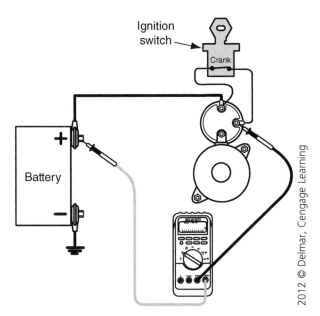

Ignition switch

Crank

Battery

2012 © Delmar, Cengage Learning

2. In the above illustration, a voltage drop test of the starter motor control circuit is being performed. With the ignition disabled, the engine is cranked over with the voltmeter on the lowest scale. Technician A says that when a reading above 1 volt is obtained, individual components in the circuit will need to be tested to find the problem. Technician B says that a reading below 3.5 volts is OK. Who is correct?

TASK A.19

A. A only
B. B only
C. Both A and B
D. Neither A nor B

Answer A is correct. Only Technician A is correct. This test is an accumulative total of all voltage drops in the circuit. There are several connections and components involved. Each will need to be tested individually to pinpoint the fault.

Answer B is incorrect. A 3.5 volt drop in a circuit is excessive, and a problem exists. The voltmeter leads must be placed across each section of the circuit and the test repeated until the high resistance is found.

Answer C is incorrect. Only Technician A is correct.

Answer D is incorrect. Technician A is correct.

3. The ignition control module uses a digital signal from the PCM (powertrain control module) for what purpose?

TASK B.7

A. RPM signal
B. Cylinder ID
C. Timing control
D. Synchronization signal

Answer A is incorrect. The ignition control module does not need the RPM signal.

Answer B is incorrect. The cam sensor is responsible for the Cylinder ID or Sync signal; the ignition control module does not need this, however.

Answer C is correct. The PCM sends a digital signal to the ignition control module to control the ignition timing and timing advance.

Answer D is incorrect. The ignition control module does not need a synchronization signal.

4. Technician A says a small amount of anti-seize should be used when replacing an oxygen sensor. Technician B says zirconium oxygen sensors require the use of thread locker when replacing. Who is correct?

    A. A only
    B. B only
    C. Both A and B
    D. Neither A nor B

    Answer A is correct. Only Technician A is correct. Anytime the oxygen sensor is removed and reinstalled or a new one is installed, a small amount of anti-seize should be put on the threads; many new oxygen sensors come with anti-seize already on the threads.

    Answer B is incorrect. Regardless of the type of oxygen sensor or air/fuel ratio sensor, anti-seize should be used to help prevent the sensor from seizing up in the manifold or exhaust.

    Answer C is incorrect. Only Technician A is correct.

    Answer D is incorrect. Technician A is correct.

5. The plunger of a PCV valve is stuck in the maximum flow position. Technician A says this can cause a rough idle. Technician B says this could cause excessive oil consumption. Who is correct?

    A. A only
    B. B only
    C. Both A and B
    D. Neither A nor B

    Answer A is incorrect. Technician B is also correct.

    Answer B is incorrect. Technician A is also correct.

    Answer C is correct. Both Technicians are correct. A PCV valve is a calculated vacuum leak. If the plunger were to stick in the maximum flow position, then the engine RPM can become high, or the engine may idle roughly. With the valve stuck in the maximum flow position, excessive oil could be pulled in the valve causing an oil consumption complaint.

    Answer D is incorrect. Both Technicians are correct.

6. A vehicle with a no-start condition in being diagnosed. The vehicle has secondary ignition voltage at the spark plug end, but no fuel pressure. The vehicle registers 12 volts at the fuel pump connector at key on and while cranking, but the fuel pump will not operate. Technician A says the fuel pump could be bad and might need to be replaced. Technician B says the positive and ground side of the fuel pump should be checked for voltage drop first. Who is correct?

    A. A only
    B. B only
    C. Both A and B
    D. Neither A nor B

    Answer A is incorrect. Technician B is also correct.

    Answer B is incorrect. Technician A is also correct.

    Answer C is correct. Both Technicians are correct. The fuel pump could be defective, but the fuel pump should not be replaced until a voltage drop test is performed on the positive and ground side of the fuel pump.

    Answer D is incorrect. Both Technicians are correct.

7. Regardless of sensor type, what attribute of a speed sensor signal does the PCM process to determine the vehicle speed?

TASK E.4

   A. Amplitude
   B. Voltage
   C. Frequency
   D. Pulse width.

   Answer A is incorrect. The amplitude is the maximum deviation of an alternating current from its average value.

   Answer B is incorrect. Speed sensors are usually magnetic pulse generators or Hall-effect. The change in voltage is not what the PCM looks at.

   Answer C is correct. The frequency is what determines vehicle speed.

   Answer D is incorrect. The speed sensor is not a pulse width producing component.

8. The voltage signal of a negative temperature coefficient sensor will do what?

TASK E.4

   A. Decrease as temperature goes up
   B. Increase as temperature goes up
   C. Be around 12 volts on most applications
   D. Be low if the sensor is open

   Answer A is correct. As temperature goes up, resistance goes down, which will cause the voltage to increase.

   Answer B is incorrect. This is describing a positive temperature coefficient sensor.

   Answer C is incorrect. Most temperature sensors operate on a 5-volt reference.

   Answer D is incorrect. If the sensor becomes open, then the voltage signal will be high, around 5 volts.

9. Technician A says that a defective starter drive gear can cause a starter to have a whining noise. Technician B says a defective starter solenoid can cause a starter to whine. Who is correct?

TASK A.22

   A. A only
   B. B only
   C. Both A and B
   D. Neither A nor B

   Answer A is incorrect. Technician B is also correct.

   Answer B is incorrect. Technician A is also correct.

   Answer C is correct. Both Technicians are correct. A defective starter drive gear can cause the starter to whine due to poor gear engagement. If the starter solenoid fails to fully engage the starter drive, however, a whining noise can also be caused.

   Answer D is incorrect. Both Technicians are correct.

TASK A.22

10. A vehicle with a manual transmission will not crank. Technician A says the neutral safety switch might be stuck open. Technician B says the starter solenoid might be stuck closed. Who is correct?

   A. A only

   B. B only

   C. Both A and B

   D. Neither A nor B

Answer A is incorrect. A manual transmission has a clutch safety switch, not a neutral safety switch.

Answer B is incorrect. A stuck closed starter solenoid would cause the starter to operate without disengaging.

Answer C is incorrect. Neither Technician is correct.

Answer D is correct. Neither Technician is correct. The neutral safety switch is open any time the transmission is in a gear to prevent starter engagement in gear. Manual transmission vehicles do not use neutral safety switches, they use clutch safety switches. The starter solenoid is opened until the ignition switch is moved to the start position, and then the starter solenoid is closed causing the starter to engage.

TASK B.3

11. An ammeter is used on the ignition primary circuit to check what?

   A. Resistance

   B. Voltage

   C. Current draw

   D. Polarity

Answer A is incorrect. An ohmmeter is used to check resistance.

Answer B is incorrect. A voltmeter is used to check voltage.

Answer C is correct. An ammeter is used to check primary ignition current draw.

Answer D is incorrect. An oscilloscope is used to check for correct polarity.

TASK C.14

12. A turbocharger fails frequently due to bearing failure. Technician A says oil contamination from turbo boost buildup in the engine could be the cause. Technician B says restricted oil passages to the turbocharger could be the problem. Who is correct?

   A. A only

   B. B only

   C. Both A and B

   D. Neither A nor B

Answer A is incorrect. Boost pressures do not contaminate the oil.

Answer B is correct. Only Technician B is correct. Sludge buildup in the oil passages can cause the passages to restrict oil flow, which would cause premature bearing failure.

Answer C is incorrect. Only Technician B is correct.

Answer D is incorrect. Technician B is correct.

13. Which of the following is the first noticeable symptom of a restricted air filter?

    A. Loss of power
    B. Decrease in fuel economy
    C. A no-start condition
    D. Excessive oil consumption

    TASK C.9

    Answer A is incorrect. If the air filter gets dirty enough, then a loss of power would exist but would not be the first thing noticed.

    Answer B is correct. The customer would complain of poor fuel economy first.

    Answer C is incorrect. If the air filter gets dirty enough, then it may cause a no-start but would not be the first thing noticed.

    Answer D is incorrect. A dirty air filter will not cause oil excessive consumption.

14. Technician A says to always check the exhaust passages when replacing an exhaust gas recirculation (EGR) valve. Technician B says an EGR valve lowers carbon monoxide emissions. Who is correct?

    A. A only
    B. B only
    C. Both A and B
    D. Neither A nor B

    TASK D.2.3

    Answer A is correct. Only Technician A is correct. EGR passages become restricted with mileage. It is always necessary to check the condition of the EGR exhaust passages when replacing the EGR valve.

    Answer B is incorrect. The purpose of the EGR system is to lower the formation of oxides of nitrogen in the exhaust.

    Answer C is incorrect. Only Technician A is correct.

    Answer D is incorrect. Technician A is correct.

15. Technician A says when completing monitor readiness, the vehicle must be driven under certain specific conditions in order for the monitors to run. Technician B says the emissions system that is being monitored must not have any related DTCs (diagnostic trouble codes) stored. Who is correct?

    A. A only
    B. B only
    C. Both A and B
    D. Neither A nor B

    TASK E.1

    Answer A is incorrect. Technician B is also correct.

    Answer B is incorrect. Technician A is also correct.

    Answer C is correct. Both Technicians are correct. Before the monitors can be run on certain systems, the vehicle must be driven under certain conditions. The service manual has the particular conditions for each monitor. If a DTC is stored for the system being checked, then the monitor will not run until the fault is repaired.

    Answer D is incorrect. Both Technicians are correct.

TASK A.10

16. A 5-gas exhaust emissions analyzer may be used to help diagnose all of the following problems EXCEPT:

    A. Cylinder misfire

    B. Stuck-open injector

    C. Burnt valve

    D. Lazy oxygen sensor

    Answer A is incorrect. A cylinder misfire will cause the hydrocarbons to be high and the oxygen to be high.

    Answer B is incorrect. A stuck-open injector will cause the carbon monoxide and hydrocarbons to be high.

    Answer C is incorrect. A burnt valve will cause the hydrocarbons to be high.

    Answer D is correct. A lazy oxygen sensor is working; it is just working slowly. An oscilloscope is needed to diagnose a lazy oxygen sensor.

TASK B.4

17. When inspecting spark plugs, which one of the following would indicate a detonation problem?

    A. White insulator tip

    B. Wet carbon deposits on the insulator tip

    C. Black insulator tip

    D. Broken insulator tip

    Answer A is incorrect. A white insulator tip is an indication of a good clean running engine.

    Answer B is incorrect. Wet carbon deposits are an indication of oil fouling.

    Answer C is incorrect. A black insulator tip indicates a rich condition.

    Answer D is correct. Detonation is an uncontrolled explosion in the combustion chamber and, left unattended, will cause holes in pistons. One sign of detonation is a cracked or broken spark plug insulator tip.

TASK C.1

18. Which of the following is the LEAST LIKELY cause of poor fuel mileage on a vehicle with SFI?

    A. Restricted fuel pressure regulator

    B. Fuel pressure regulator partially stuck open

    C. Injector stuck open

    D. Burnt valve

    Answer A is incorrect. A restricted fuel pressure regulator will cause high fuel pressures that will result in a loss of fuel mileage.

    Answer B is correct. A partially stuck-open pressure regulator will cause low fuel pressure but will not cause a loss in fuel mileage.

    Answer C is incorrect. A stuck-open fuel injector will cause a loss of fuel mileage.

    Answer D is incorrect. A burnt valve will cause a loss of fuel mileage.

19. Which of the follow is a function of an AIR system diverter valve?

    A. Lean out the mixture on deceleration.

    B. Richen the mixture on deceleration.

    C. Prevent a backfire on deceleration.

    D. Pull fresh air into the exhaust on deceleration.

TASK D.3.3

Answer A is incorrect. The PCM leans out the mixture on deceleration my decreasing the injector pulse width.

Answer B is incorrect. The mixture is not richened on deceleration. It is leaned out, but not by the AIR system diverter valve.

Answer C is correct. On deceleration the secondary air is diverted to the air cleaner to prevent an exhaust backfire.

Answer D is incorrect. If fresh air were pulled into the exhaust on deceleration, the vehicle would backfire out the exhaust.

20. Which of the following computer sensor signals is not used during open-loop mode?

    A. Manifold absolute pressure (MAP) sensor

    B. Throttle position sensor (TPS)

    C. Engine coolant temperature sensor

    D. Heated O_2 sensor

TASK E.4

Answer A is incorrect. The MAP sensor is used during open loop.

Answer B is incorrect. The TPS (throttle position sensor) is used during open loop.

Answer C is incorrect. The ECT is used during open loop.

Answer D is correct. The heated O_2 sensor is ignored during open loop. When the vehicle enters closed loop, the heated O_2 sensor is read by the PCM to control the air/fuel mixture.

21. Which of the following is the LEAST LIKELY cause of low cylinder compression?

    A. Burnt valve

    B. Worn valve guides

    C. Worn piston rings

    D. Worn valve seats

TASK A.7

Answer A is incorrect. A bunt valve will cause low engine compression.

Answer B is correct. A worn valve guide can make the engine smoke but will not affect engine compression.

Answer C is incorrect. Worn piston rings will cause low engine compression.

Answer D is incorrect. Worn valve seats will cause engine compression to be low.

TASK A.18

22. A starter free-spinning test is being done on the bench with a fully charged battery. The current draw is higher than specification and the RPM is lower. Technician A says this could be caused by tight bushings. Technician B says typical current draw when free spinning a starter is 60 to 100 amps. Who is correct?

    A. A only
    B. B only
    C. Both A and B
    D. Neither A nor B

    Answer A is incorrect. Technician B is also correct.

    Answer B is incorrect. Technician A is also correct.

    Answer C is correct. Both Technicians are correct. Every starter should be bench tested before installing on the vehicle. Most starters will have a current draw of 60 to 100 amps when free spinning on the bench. If the current draw is too high and the starter turns slowly, then the bushings could be too tight.

    Answer D is incorrect. Both Technicians are correct.

TASK B.4

23. The secondary ignition system can be tested using which device?

    A. Test light
    B. Ohmmeter
    C. An ammeter
    D. Logic probe

    Answer A is incorrect. A test light can be used to test the primary ignition, but not the secondary ignition system.

    Answer B is correct. An ohmmeter can be used to measure the resistance of secondary ignition components.

    Answer C is incorrect. The secondary ignition system puts out very low amperage, and amperage is not used for diagnostic purposes.

    Answer D is incorrect. A logic probe is used to test delicate computer circuits, but not high voltage secondary circuits.

TASK C.5

24. Which of these steps should the technician take first when faulty fuel pump pressure is suspected?

    A. Check the fuel filter for restriction
    B. Check for kinked fuel lines
    C. Check fuel pressure and volume
    D. Check for fuel related diagnostic trouble codes (DTCs)

    Answer A is incorrect. Checking for filter restrictions comes after the system pressure and volume are checked.

    Answer B is incorrect. Fuel lines should be checked for kinks, but after the fuel pressure and volume are checked.

    Answer C is correct. The first step when a faulty fuel pump pressure is suspected is to check the pressure and volume of the fuel pump.

    Answer D is incorrect. Checking for fuel-related DTCs comes after the pressure and volume are checked.

25. What should the fuel level be before an evaporative emissions monitor will run?

    A. 50 percent
    B. Over 25 percent
    C. At least 75 percent
    D. Between 15 percent and 85 percent

TASK D.4.1

Answer A is incorrect. The fuel level must be between 15 percent and 85 percent for the Evaporative emissions monitor to run.

Answer B is incorrect. The fuel level must be between 15 percent and 85 percent for the Evaporative emissions monitor to run.

Answer C is incorrect. The fuel level must be between 15 percent and 85 percent for the Evaporative emissions monitor to run.

Answer D is correct. The fuel level must be between 15 percent and 85 percent for the Evaporative emissions monitor to run.

26. A vehicle has a misfire DTC. Technician A says the PCM uses the camshaft sensor signal to monitor engine misfire. Technician B says a misfire diagnostic code is a high-priority code and should be diagnosed first before other non-emissions-related codes. Who is correct?

    A. A only
    B. B only
    C. Both A and B
    D. Neither A nor B

TASK E.2

Answer A is incorrect. The PCM uses the crank shaft sensor signal for misfire detection.

Answer B is correct. Only Technician B is correct. A misfire code is a high-priority code because the catalytic converter can be destroyed by a misfire, and this type of code should be the first to be diagnosed and repaired.

Answer C is incorrect. Only Technician B is correct.

Answer D is incorrect. Technician B is correct.

27. Technician A says catalytic converter efficiency is not monitored on OBD-II compliant vehicles. Technician B says some vehicles may have no computer diagnostic capabilities for the secondary air-injection system. Who is correct?

    A. A only
    B. B only
    C. Both A and B
    D. Neither A nor B

TASK D.3.2

Answer A is incorrect. Technician A is incorrect.

Answer B is correct. Technician B is correct. Some vehicles may not have computer diagnostic capabilities for the secondary air-injection system. Consult the service manual for information.

Answer C is incorrect. Only Technician B is correct.

Answer D is incorrect. Technician B is correct.

TASK A.3

28. A vehicle with double overhead valves is being diagnosed for backfiring through the exhaust manifold. Technician A says this engine uses a separate camshaft for the intake and exhaust valves. Technician B says a 4-cylinder DOHC (double overhead camshaft) engine has two camshafts. Who is correct?

A. A only

B. B only

C. Both A and B

D. Neither A nor B

Answer A is incorrect. Technician B is also correct.

Answer B is incorrect. Technician A is also correct.

Answer C is correct. Both Technicians are correct. An engine with double overhead valves will have a cam shaft for the intake valves and a camshaft for the exhaust valves. A 4-cylinder engine will have two camshafts; a V6 DOHC engine will have four camshafts.

Answer D is incorrect. Both Technicians are correct.

TASK A.17

29. A vehicle with a hard-starting complaint is being diagnosed. The battery shows an open circuit voltage of 11.7 volts. Technician A says the battery should be recharged then load-tested. Technician B says the battery cranking voltage should also be checked. Who is correct?

A. A only

B. B only

C. Both A and B

D. Neither A nor B

Answer A is incorrect. Technician B is also correct.

Answer B is incorrect. Technician A is also correct.

Answer C is correct. Both Technicians are correct. An open circuit voltage of 11.7 volts is too low; the battery should be recharged then load-tested. Another test that could be done is a cranking voltage test.

Answer D is incorrect. Both Technicians are correct.

TASK B.4

30. High secondary ignition-system circuit resistance can be caused by all of the following EXCEPT:

A. Excessive spark plug gap

B. Spark plug gap bridging

C. Open circuit in the ignition coil

D. Excessive distributor cap terminal to rotor gap

Answer A is incorrect. Excessive spark plug gap increases secondary resistance.

Answer B is correct. Spark plug gap bridging decreases secondary resistance.

Answer C is incorrect. An open in the ignition coil secondary windings is a cause of high secondary resistance.

Answer D is incorrect. Excessive distributor cap terminal to rotor gap increases secondary resistance.

31. A vacuum assisted fuel pressure regulator is used on sequential fuel injection (SFI) for which of the following reasons?

    A. To increase fuel delivery under high load conditions
    B. To provide a constant pressure drop across the injector due to a rapidly opening throttle
    C. To improve injector spray patterns
    D. To prevent fuel-pressure leak down when the engine is turned off

TASK C.6

    Answer A is incorrect. Fuel delivery is increased by increasing the injector pulse width.

    Answer B is correct. If the throttle is opened rapidly, then a fuel pressure drop can be caused by all the injectors opening at once. The vacuum assist helps get the pressure drop constant.

    Answer C is incorrect. Injector spray pattern is controlled by the tip of the injector.

    Answer D is incorrect. Leak down should not exist on a fuel system, but the vacuum assist does not prevent it.

32. Technician A says some vehicles illuminate a loose gas cap light and store a code if a very large leak is detected. Technician B says the vent valve is a normally closed solenoid. Who is correct?

    A. A only
    B. B only
    C. Both A and B
    D. Neither A nor B

TASK D.4.3

    Answer A is correct. Only Technician A is correct. On some vehicles, if a very large leak is detected, then a loose gas cap light is illuminated and a DTC stored in the PCM.

    Answer B is incorrect. The purge control solenoid is a normally closed solenoid, and the vent solenoid is normally open.

    Answer C is incorrect. Only Technician A is correct.

    Answer D is incorrect. Technician A is correct.

33. The cause of a MIL (malfunction indicator lamp) not illuminating during a bulb check when the ignition is turned on can be all of the following EXCEPT:

    A. A blown bulb
    B. An internal circuit problem in the PCM
    C. A open in the MIL circuit
    D. A DTC stored in the PCM

TASK E.3

    Answer A is incorrect. A blown bulb will prevent the bulb from illuminating during a bulb check.

    Answer B is incorrect. The PCM controls the MIL. An internal problem in the PCM can cause the MIL not to illuminate during a bulb check.

    Answer C is incorrect. If the MIL circuit becomes open, then the MIL will not illuminate during a bulb check.

    Answer D is correct. A stored DTC would cause the MIL to illuminate.

TASK A.3

34. Which of the following is the LEAST LIKELY cause of blue exhaust smoke from the engine?

    A. Worn piston rings
    B. Worn valve guides
    C. Worn valve seats
    D. Worn valve seals

    Answer A is incorrect. Worn piston oil control rings will cause blue smoke.

    Answer B is incorrect. Worn valve guides will cause blue smoke, especially at start up.

    Answer C is correct. Worn valve sets will cause a miss, not blue smoke.

    Answer D is incorrect. Worn valve seals will cause blue smoke at start up.

TASK B.7

35. Technician A says that the ignition module controls the timing during start up on some ignition systems. Technician B says that on some ignition systems, the PCM has full control of timing at all times. Who is correct?

    A. A only
    B. B only
    C. Both A and B
    D. Neither A nor B

    Answer A is incorrect. Technician B is also correct.

    Answer B is incorrect. Technician A is also correct.

    Answer C is correct. Both Technicians are correct. Many vehicles' ignition timing is controlled by the ignition module on startup, which is called *module* or *base timing mode*. After a specific time, the PCM takes over the timing for more precise timing control. Some vehicles do not use a module mode-base timing; instead, the PCM has full control all the time.

    Answer D is incorrect. Both Technicians are correct.

TASK C.8

36. A vehicle with SFI is running rough. A lab scope shows all injector waveforms to be identical except for one that has a considerably shorter voltage spike than the others. Which of the following is the most likely cause?

    A. Open connection at the injector
    B. Bad PCM
    C. Shorted injector winding
    D. Low charging system voltage

    Answer A is incorrect. An open connection would not have any pattern.

    Answer B is incorrect. PCM injector drivers either will or will not work; they usually do not cause a short pattern.

    Answer C is correct. An injector is a solenoid, when the PCM turns off the injector the magnetic field collapses and is displayed as a voltage spike. A low spike indicates a shorted injector winding.

    Answer D is incorrect. Low charging system voltage would affect all injectors.

37.   Technician A says if the EGR valve remains open at idle and low speed, then detonation can occur. Technician B says if the EGR valve does not open at cruising speeds, then the vehicle will run rough. Who is correct?

TASK D.2.1

   A.   A only

   B.   B only

   C.   Both A and B

   D.   Neither A nor B

Answer A is incorrect. If the EGR valve opens at idle, then the vehicle will run rough or stall at idle.

Answer B is incorrect. If the EGR does not open at cruising speeds, then the vehicle will have spark knock and detonation.

Answer C is incorrect. Neither Technician is correct.

Answer D is correct. Neither Technician is correct. The EGR valve should be closed at idle and open at cruising speeds. If the EGR valve opens at idle, then the vehicle will run rough and possibly stall. If the EGR valve fails to open at cruising speeds, then the vehicle can have detonation and spark knock.

38.   A soft code in the PCM memory is a code that:

TASK E. 1

   A.   Indicates a top priority code

   B.   Occurred in the past, but no longer exists

   C.   Exists at the time the vehicle is being tested

   D.   Can be found if a diagnostic flow chart is used

Answer A is incorrect. A soft code has nothing to do with priority.

Answer B is correct. A soft code is an intermittent fault that no longer exists.

Answer C is incorrect. A hard fault is one that exists at the time the vehicle is being tested.

Answer D is incorrect. Soft faults are hard to diagnose because a diagnostic flow chart will not help if the problem is not occurring at the time of the test.

39.   A vehicle with a slow cranking complaint has a voltmeter connected to the 12-volt battery. With the engine cranking, what is the lowest recommended voltmeter reading?

TASK A.17

   A.   9.6 volts

   B.   11.0 volts

   C.   7.5 volts

   D.   12.0 volts

Answer A is correct. The voltage should never drop below 9.6 volts.

Answer B is incorrect. It is good if the voltage stays at 11.0 volts, but that is not at the minimum allowed.

Answer C is incorrect. 7.5 volts is too low; 9.6 volts is the minimum.

Answer D is incorrect. It is good if the voltage stays at 12.0 volts, but that is not at the minimum allowed.

TASK B.6

40. Technician A says the crankshaft sensor may be rotated to adjust the base ignition timing on some engines with electronic ignition. Technician B says on some systems, the crankshaft sensor interrupter ring is part of the crankshaft. Who is correct?

   A. A only
   B. B only
   C. Both A and B
   D. Neither A nor B

Answer A is incorrect. Technician B is also correct.

Answer B is incorrect. Technician A is also correct.

Answer C is correct. Both Technicians are correct. Some vehicles have adjustable crankshaft sensors to allow timing adjustments. Some vehicles have the crankshaft sensor interrupter ring forged into the crankshaft, and the sensor goes through the block.

Answer D is incorrect. Both Technicians are correct.

TASK C.9

41. Technician A says some vehicles are more sensitive to aftermarket high-performance air filters than others. Technician B says to always check technical service bulletins when an aftermarket high-performance air filter has been installed and a drivability complaint is being diagnosed. Who is correct?

   A. A only
   B. B only
   C. Both A and B
   D. Neither A nor B

Answer A is incorrect. Technician B is also correct.

Answer B is incorrect. Technician A is also correct.

Answer C is correct. Both Technicians are correct. Many vehicles will experience drivability complaints due to the adding of a high-performance air filter. Some manufacturers have issued technical service bulletins dealing with this issue.

Answer D is incorrect. Both Technicians are correct.

TASK D.3.3

42. While testing a pulse-air-type air-injection system, Technician A says with the air cleaner's fresh-air hose removed and the engine idling, there should be steady audible pulses at the air cleaner inlet. Technician B says on some vehicles, air pulses should be felt escaping through the air cleaner inlet with the fresh air hose removed. Who is correct?

   A. A only
   B. B only
   C. Both A and B
   D. Neither A nor B

Answer A is correct. Only Technician A is correct. Pulse air-injection systems can be noisy, especially if the fresh air hose is left off. With the hose off, the pulse air-injection system makes a steady audible noise.

Answer B is incorrect. The pulses of the pulse air-injection system cannot be felt at the fresh air inlet.

Answer C is incorrect. Only Technician A is correct.

Answer D is incorrect. Technician A is correct.

43. Technician A says freeze frame data is always set for a misfire code. Technician B says a misfire code is considered a high-priority code. Who is correct?

TASK E.1

   A. A only
   B. B only
   C. Both A and B
   D. Neither A nor B

   Answer A is incorrect. Technician B is also correct.

   Answer B is incorrect. Technician A is also correct.

   Answer C is correct. Both Technicians are correct. Any time a misfire code is set, a freeze frame data will be stored with the code. A misfire code is a high-priority code because the catalytic converter can be destroyed if left unattended.

   Answer D is incorrect. Both Technicians are correct.

44. An oxygen sensor reads higher-than-normal voltage. Technician A says that the engine may have a vacuum leak. Technician B says that the exhaust manifold could be leaking. Who is correct?

TASK E.4

   A. A only
   B. B only
   C. Both A and B
   D. Neither A nor B

   Answer A is incorrect. A vacuum leak would cause the oxygen sensor to read lower-than-normal voltage.

   Answer B is incorrect. A leaking exhaust manifold would cause the oxygen sensor to read lower-than-normal voltage.

   Answer C is incorrect. Neither Technician is correct.

   Answer D is correct. Neither Technician is correct. The oxygen sensor produces a voltage from 100 millivolts to 900 millivolts. A vacuum leak or leaking exhaust manifold would cause the voltage to be under 450 millivolts.

45. A fully charged 12-volt battery should measure what voltage?

TASK A.17

   A. 12.0 volts
   B. 12.3 volts
   C. 12.5 volts
   D. 12.6 volts

   Answer A is incorrect. 12 volts is too low.

   Answer B is incorrect. 12.3 volts is too low.

   Answer C is incorrect. 12.5 volts is too low.

   Answer D is correct. A fully charged battery should measure 12.6 volts with no load applied.

TASK B.5

46. An open spark plug wire was found on an engine that was missing on acceleration. Technician A says that the distributor cap and rotor should be carefully inspected for carbon tracks. Technician B says that the ignition coil should be replaced because the open wire could have caused the coil to become tracked internally. Who is correct?

   A.   A only
   B.   B only
   C.   Both A and B
   D.   Neither A nor B

   Answer A is incorrect. Technician B is also correct.

   Answer B is incorrect. Technician A is also correct.

   Answer C is correct. Both Technicians are correct. An open spark plug wire allows excessively high secondary voltage spikes. These spikes can damage the distributor cap, rotor, and even cause the ignition coil to become tracked internally. After inspecting the cap and rotor, the ignition coil should be replaced because of the possibility of being internally tracked.

   Answer D is incorrect. Both Technicians are correct.

TASK C.8

47. Using which of the following can an injector pulse can be tested at the injector connector?

   A.   Neon spark tester
   B.   Vacuum hose and a test light
   C.   DVOM
   D.   NOID light

   Answer A is incorrect. A spark tester is not used to check for injector pulse.

   Answer B is incorrect. A vacuum hose and test light is used to perform a power balance test.

   Answer C is incorrect. A DVOM is not an accurate method of checking for injector pulse.

   Answer D is correct. A NOID light is used to check for injector pulses.

TASK D.4.1

48. An evaporative emissions system that does not purge correctly can cause all of the following EXCEPT:

   A.   Loss of fuel mileage
   B.   Increased crankcase blowby
   C.   Rich air/fuel ration
   D.   An increase in tail pipe emissions

   Answer A is incorrect. If purging takes place at the wrong time, then a loss of fuel economy or mileage can take place.

   Answer B is correct. The EVAP system purging at the wrong time will not cause increased crankcase blowby.

   Answer C is incorrect. Purging at the wrong time will cause a rich air/fuel ratio.

   Answer D is incorrect. If the EVAP system purges at the wrong time, then tail pipe emissions will be increased.

49. A short-term fuel trim of 0 percent and a long-term fuel trim of +19 percent means what?

TASK E.3

   A. The engine has a history of running rich.
   B. The engine is running lean at the present time.
   C. The engine is running rich at the present time.
   D. The engine has a history of running lean.

   Answer A is incorrect. If the engine had a history of running rich, then the long-term would be -19 percent.

   Answer B is incorrect. The short-term fuel trim is 0 percent, so no correction is being made at present time.

   Answer C is incorrect. The short-term fuel trim is 0 percent, so no correction is being made at present time.

   Answer D is correct. The long-term fuel trim is +19 percent, meaning the PCM is adding 19 percent fuel to correct the vehicle's lean condition history.

50. A typical oxygen sensor output signal can be measured with the DVOM set on what unit of measure?

TASK E.4

   A. AC volts
   B. DC volts
   C. Ohms
   D. Frequency

   Answer A is incorrect. The oxygen sensor produces a DC voltage signal.

   Answer B is correct. The oxygen sensor produces a DC voltage signal.

   Answer C is incorrect. The oxygen sensor produces a DC voltage signal.

   Answer D is incorrect. The oxygen sensor produces a DC voltage signal.

PREPARATION EXAM ANSWER SHEET FORMS

ANSWER SHEET

1. _____	21. _____	41. _____
2. _____	22. _____	42. _____
3. _____	23. _____	43. _____
4. _____	24. _____	44. _____
5. _____	25. _____	45. _____
6. _____	26. _____	46. _____
7. _____	27. _____	47. _____
8. _____	28. _____	48. _____
9. _____	29. _____	49. _____
10. _____	30. _____	50. _____
11. _____	31. _____	
12. _____	32. _____	
13. _____	33. _____	
14. _____	34. _____	
15. _____	35. _____	
16. _____	36. _____	
17. _____	37. _____	
18. _____	38. _____	
19. _____	39. _____	
20. _____	40. _____	

ANSWER SHEET

1. _____	21. _____	41. _____
2. _____	22. _____	42. _____
3. _____	23. _____	43. _____
4. _____	24. _____	44. _____
5. _____	25. _____	45. _____
6. _____	26. _____	46. _____
7. _____	27. _____	47. _____
8. _____	28. _____	48. _____
9. _____	29. _____	49. _____
10. _____	30. _____	50. _____
11. _____	31. _____	
12. _____	32. _____	
13. _____	33. _____	
14. _____	34. _____	
15. _____	35. _____	
16. _____	36. _____	
17. _____	37. _____	
18. _____	38. _____	
19. _____	39. _____	
20. _____	40. _____	

ANSWER SHEET

1. _____	21. _____	41. _____
2. _____	22. _____	42. _____
3. _____	23. _____	43. _____
4. _____	24. _____	44. _____
5. _____	25. _____	45. _____
6. _____	26. _____	46. _____
7. _____	27. _____	47. _____
8. _____	28. _____	48. _____
9. _____	29. _____	49. _____
10. _____	30. _____	50. _____
11. _____	31. _____	
12. _____	32. _____	
13. _____	33. _____	
14. _____	34. _____	
15. _____	35. _____	
16. _____	36. _____	
17. _____	37. _____	
18. _____	38. _____	
19. _____	39. _____	
20. _____	40. _____	

ANSWER SHEET

1. _____	21. _____	41. _____
2. _____	22. _____	42. _____
3. _____	23. _____	43. _____
4. _____	24. _____	44. _____
5. _____	25. _____	45. _____
6. _____	26. _____	46. _____
7. _____	27. _____	47. _____
8. _____	28. _____	48. _____
9. _____	29. _____	49. _____
10. _____	30. _____	50. _____
11. _____	31. _____	
12. _____	32. _____	
13. _____	33. _____	
14. _____	34. _____	
15. _____	35. _____	
16. _____	36. _____	
17. _____	37. _____	
18. _____	38. _____	
19. _____	39. _____	
20. _____	40. _____	

ANSWER SHEET

1. _____	21. _____	41. _____
2. _____	22. _____	42. _____
3. _____	23. _____	43. _____
4. _____	24. _____	44. _____
5. _____	25. _____	45. _____
6. _____	26. _____	46. _____
7. _____	27. _____	47. _____
8. _____	28. _____	48. _____
9. _____	29. _____	49. _____
10. _____	30. _____	50. _____
11. _____	31. _____	
12. _____	32. _____	
13. _____	33. _____	
14. _____	34. _____	
15. _____	35. _____	
16. _____	36. _____	
17. _____	37. _____	
18. _____	38. _____	
19. _____	39. _____	
20. _____	40. _____	

ANSWER SHEET

1. _____

2. _____

3. _____

4. _____

5. _____

6. _____

7. _____

8. _____

9. _____

10. _____

11. _____

12. _____

13. _____

14. _____

15. _____

16. _____

17. _____

18. _____

19. _____

20. _____

21. _____

22. _____

23. _____

24. _____

25. _____

26. _____

27. _____

28. _____

29. _____

30. _____

31. _____

32. _____

33. _____

34. _____

35. _____

36. _____

37. _____

38. _____

39. _____

40. _____

41. _____

42. _____

43. _____

44. _____

45. _____

46. _____

47. _____

48. _____

49. _____

50. _____

Glossary

Accelerator A control, usually foot-operated, linked to the throttle valve of the carburetor.

Accelerator Pump A pump in the carburetor that generates additional fuel to cover for transitions that occur when the throttle position is changed.

Accessory Drive As in the belt driven accessories under the hood—fan, alternator, A/C, power steering, air-injection pump.

Air/Fuel Mixture The proportion of air and fuel supplied to the engine.

Analyzer Any device, such as an oscilloscope, having readout provisions used to troubleshoot a function or event as an aid in making proper repairs.

Automatic Choke A system that positions the choke automatically.

Back Pressure The excessive pressure buildup in an engine crankcase; the resistance of an exhaust system.

Battery A device used to store electrical energy in chemical form.

Battery Cable Heavy wires connected to the battery for positive (hot) and negative (ground) leads.

Battery Charger A device used to charge and recharge a battery.

Bearing A device having an inner and outer race with one or more rows of steel balls.

Bottom Dead Center (BDC) when the piston is at the bottom of the cylinder.

Catalytic Converter An exhaust system component designed to reduce oxides of nitrogen (NOx), hydrocarbon (HC), and carbon monoxide (CO).

Check Valve A device that allows the flow of liquid or vapor in one direction and blocks it in the other direction.

Coil That part of the ignition system that provides high voltage for the spark plugs.

Cold-Cranking Amperage The number of amperes that a fully charged battery will provide for 30 seconds without the terminal voltage dropping below 7.2 volts.

Combustion Chamber The area above a piston, at top dead center, where combustion takes place.

Compression The process of squeezing a vapor into a smaller space.

Compression Test A measurement of the pressure a cylinder is able to generate during a controlled cranking period.

Computer An electronic device for storing and processing data, typically in binary form, according to instructions given to it in a variable program.

Cooling System The radiator, hoses, heater core, and cooling jackets used to carry away engine heat and dissipate it in the surrounding air.

Cruise Control A system of automatically maintaining preset vehicle speed over varying terrain.

Customer Complaint The description of a problem provided by the customer, usually the driver of the vehicle.

Cylinder Balance A dynamic test that shorts out the engine cylinders one at a time and compares the power loss in each to pinpoint weak cylinders.

Cylinder Head That part that covers the cylinders and pistons.

Cylinder Leakage Test A test to determine how well a cylinder seals when the piston is at top dead center and the valves are closed. Also known as a leak down test.

Deck The flat mating surfaces of an engine block and head.

Dedicated Ground A grounded connection dedicated to a particular component or circuit on an automobile.

Diaphragm A flexible, rubber-like membrane.

Digital Ohmmeter A device that sends a small amount of current into an isolated circuit and indicates the amount of resistance in a numerical readout.

Digital Voltmeter A device that reads the difference in voltage pressure at two points of an electrical circuit in a numerical readout.

Driveability A term used for any problem or compliant the driver might encounter in the engine control system or transmission control system.

Driveability Problem A problem or compliant encountered in the engine control system or transmission control system.

Drive Belt The belt or belts used to drive the engine-mounted accessories off the crankshaft.

Dwell The degree of distributor shaft rotation while the points are closed.

Early Timing A phrase used to describe advanced timing, meaning the spark is delivered to the spark plug too soon.

Electronic Control A control device that is electrically or electronically actuated.

Electronics Pertaining to that branch of science dealing with the motion, emission, and behavior of currents of free electrons.

Emission A product, harmful or not, emitted to the atmosphere. Emissions are generally regarded as harmful.

Emission Test The use of calibrated equipment to determine the amount of emissions that are being released to the atmosphere.

Emission Control (1)Emission tests are tests done on the vehicle to determine how it compares to federal government standards for tail pipe emissions, crankcase emissions and evaporative emissions. (2) Components directly or indirectly responsible for reducing harmful emissions.

Enabling Criteria Specific operating parameters that must be met before a monitor will run.

Engine A prime mover. A device for converting chemical energy (fuel) to useable mechanical energy (motion).

Engine-Management System An electronic system that monitors, regulates, and adjusts, engine performance and conditions.

Engine Manifold Vacuum The vacuum signal taken directly off the intake manifold or below the throttle plate.

Engine Oil A lubricant formulated and designated for use in an engine.

Evaporative Emissions Control (EVAP) The system that captures and stores vapors from the fuel system in a charcoal canister to be burned once the engine is started.

Exhaust (1)The byproduct of combustion. (2)The pipe from the muffler to the atmosphere.

Exhaust Back Pressure The pressure that develops in the exhaust system during normal operation. Two pounds is cause for concern.

Exhaust Gas Recirculation (EGR) System A small amount of exhaust gas sent into the intake manifold during light cruise conditions to lower combustion chamber temperatures and reduce the formation of nitrogen oxides.

Exhaust Port An opening that allows the exhaust gases to escape.

F An abbreviation for Fahrenheit, a temperature measurement in the English scale.

Fail-Safe A default mode designed into many operating systems that allows limited function of a system when a malfunction occurs. This is to protect the system or to allow the driver to move the vehicle to a safe area.

Fan Blade A flat-pitched part of a fan that moves air.

Fault Code A numeric readout system used as an aid in troubleshooting procedures, information about functions or malfunctions of an electronic system.

Fuel Contamination Any impurities in the fuel system.

Fuel Injector Electrical or mechanical devices that meters fuel into the engine.

Fuel Pressure The pressure of the fuel in an injection or non-injection fuel system.

Fuel Pressure Regulator A device that regulates fuel pressure. Fuel injected engines require pressure regulators, because some fuel pumps develop over 100 psi (690 kPa).

Fuel Pump An electrical or mechanical device that pumps fuel from the tank to the carburetor or injection system.

Fuel Volatility A term used to determine how rapidly a fuel evaporates or burns.

Green Death A term used by technicians describing the formation of green corrosion in wiring and connectors.

Hand Choke A choke that is manually controlled by cable.

Head Gasket A sealing material between the head and block.

Hot Lines Special telephone or computer lines for information access for help with problem solving.

Idle Speed The speed of an engine at idle with no load.

Ignition System The system that supplies the high voltage required to fire the spark plugs.

Ignition Timing The interval, in degrees of crankshaft rotation, before top dead center that a spark plug fires.

Intake Manifold That part of the engine that directs the air fuel mixture to the cylinders.

Intake valve A valve on the cylinder that allows the air/fuel mixture to enter the combustion chamber.

Jumped timing A term meaning valve timing is off or out of specifications; often used in reference to timing chains, belts, or gears. This happens when parts wear or loosen.

Jump Start To aid starting by the use of an external power source, such as a battery or battery charger.

Keep Alive Memory A program in many computerized devices that retains fault code information and other information necessary for the operation of the system.

Key-Off Battery Drain A term used for parasitic drains. Electrical demands on the battery while the ignition key is in the off position.

Knock Sensor A sensor that signals the engine control computer when detonation is detected retarding ignition timing.

Lean ramping A term that describes when a vehicle leans out on deceleration to the point of causing the vehicle to surge.

Long-Term Fuel Trim (LFT) "Permanent" addition or subtraction of fuel assignment for fuel injected vehicles.

Lubrication The act of applying lubricant to fittings and other moving parts.

Magnaflux A dry, nondestructive magnetic test to check for cracks or flaws in iron or steel parts.

Main Bearing The bearing that supports the crankshaft in the lower end of an engine.

Module A semiconductor device designed to control various systems like, ignition, engine control, steering, suspension, brakes, transmissions, power windows, power seats, windshield wipers, brakes, traction control, and cruise control.

Oil A lubricant.

Oscilloscope An instrument that produces a visible image of one or more rapidly varying electrical quantities with respect to time and intensity.

Oxygen Sensor A device located in the exhaust system, close to the engine, that reacts to the different amounts of oxygen present in the exhaust gases and sends signals the engine control computer so it can maintain the proper air/fuel ratio.

Parasitic Load An electrical load that is present when the ignition switch is in the off position.

Pending Conditions Circumstances that may prevent a monitor from running properly, such as an oxygen sensor fault not allowing a catalyst monitor to run.

Pings Unscheduled explosions in the combustion chamber. Also referred to as spark, knock, or detonation.

Positive Crankcase Ventilation (PCV) Valve A metering device connecting the engine crankcase to engine vacuum, which allows burning of crankcase vapors to reduce harmful engine emissions.

Power Balance A dynamic test that shorts out one engine cylinder at a time and compares the power loss to pinpoint weak cylinders.

Pounds Per Square Inch (PSI) Used in the English system for air, vacuum, and fluid pressure measurements.

Power Distribution Circuit The power and ground circuits from the battery, through the ignition switch and fuses, to the individual circuits on the vehicle.

Pulse Air-Injection System Draws outside air into the exhaust manifold by negative pressure pulses created as the exhaust is pushed out of a cylinder by the piston. Requires no power from the engine to run a pump, as with the belt-driven varieties.

Rev Limiter A device that limits the revolutions per minute of a device or component.

Revolutions Per Minute (RPM); also, rpm or r/min. The number of times a device or component spins or turns in 60 seconds.

Scanner A device, usually hand held, that accesses the electronic systems on vehicles to obtain fault codes and operating parameters. Can be used to simulate signals and to verify operation of systems on some vehicles.

Scan Tool A tester used to recall trouble codes.

Seal (1) A gasket-like material between two or more parts; (2) A ring-like gasket around a shaft to prevent fluid or vapor leakage.

Secondary Air Insertion System Outside air that is pumped into the exhaust system and the catalytic converter to promote continued emissions.

Sensor An electrical sending unit device to monitor conditions for use in controlling systems by a computer.

Serpentine Belt A wide, flat belt with multiple grooves that winds through all of the engine accessory pulleys and drives them from the load.

Shroud A hood-like device used to direct air flow.

Speed Limiter Usually a program in the engine control computer designed to limit the speed of the vehicle because of tire speed rating.

Specific Gravity The ratio of the substance to the density of water.

Supercharger A belt-driven device that pumps air into the engine induction system at a pressure much higher than atmospheric pressure.

Tail Pipe The tube-like components that direct exhaust vapors from the outlet of the muffler to the atmosphere.

Technical Information Information found in manufacturers' manuals, bulletins, reports, text books, and other such sources.

Technical Service Bulletin (TSB) Periodic information provided by the manufacturer relative to production changes and service tips.

Temperature The heat content of matter as measured on a thermometer.

Timing The spark delivery in relation to the piston position.

Top Dead Center (TDC) When the piston is at the top of the cylinder.

Trouble Code Numbers generated by the diagnostic system that refer to certain troubleshooting procedures.

Turbocharger A device, driven by exhaust gases, that pumps air into the engine induction system at a pressure higher than atmospheric pressure.

Unmetered Matter, such as air, that is entering a controlled area without being measured by a management system.

Unmetered Air Air that is entering a controlled area without being measured by the air management system.

Valve Timing The opening and closing of valves in relation to crankshaft rotation.

V-Belt A rubber-like, V-shaped belt used to drive engine-mounted accessories off the crankshaft pulley or an intermediate pulley.

Verify the Repair To retest a system and/or test drive the vehicle after repairs are made.

Voltage A quantity of electrical force.

Voltage Spikes Higher than normal voltage often caused by collapsing magnetic fields.

Wastegate A device on superchargers and turbochargers that limits the amount of pressure increase in the intake to safe design limits.

Water Pump A device, usually engine-driven, for circulating coolant in the cooling system.

Wrap A term used for the amount of contact a belt has on a pulley.

Notes

Notes

Notes

Notes